Lecture Notes in Artificial Intelligence   9918

Subseries of Lecture Notes in Computer Science

More information about this series at http://www.springer.com/series/1244

Pavel Král · Carlos Martín-Vide (Eds.)

Statistical Language and Speech Processing

4th International Conference, SLSP 2016
Pilsen, Czech Republic, October 11–12, 2016
Proceedings

 Springer

Editors
Pavel Král
University of West Bohemia
Plzeň
Czech Republic

Carlos Martín-Vide
Rovira i Virgili University
Tarragona
Spain

ISSN 0302-9743   ISSN 1611-3349 (electronic)
Lecture Notes in Artificial Intelligence
ISBN 978-3-319-45924-0   ISBN 978-3-319-45925-7 (eBook)
DOI 10.1007/978-3-319-45925-7

Library of Congress Control Number: 2016950400

LNCS Sublibrary: SL7 – Artificial Intelligence

Printed on acid-free paper

This Springer imprint is published by Springer Nature
The registered company is Springer International Publishing AG
The registered company address is: Gewerbestrasse 11, 6330 Cham, Switzerland

Preface

These proceedings contain the papers that were presented at the 4th International Conference on Statistical Language and Speech Processing (SLSP 2016), held in Pilsen, Czech Republic, during October 11–12, 2016.

SLSP deals with topics of either theoretical or applied interest, discussing the employment of statistical models (including machine learning) within language and speech processing, namely:

Anaphora and coreference resolution
Authorship identification, plagiarism, and spam filtering
Computer-aided translation
Corpora and language resources
Data mining and semantic web
Information extraction
Information retrieval
Knowledge representation and ontologies
Lexicons and dictionaries
Machine translation
Multimodal technologies
Natural language understanding
Neural representation of speech and language
Opinion mining and sentiment analysis
Parsing
Part-of-speech tagging
Question-answering systems
Semantic role labeling
Speaker identification and verification
Speech and language generation
Speech recognition
Speech synthesis
Speech transcription
Spelling correction
Spoken dialog systems
Term extraction
Text categorization
Text summarization
User modeling

SLSP 2016 received 38 submissions. Each paper was reviewed by three Program Committee members and also a few external reviewers were consulted. After a thorough and vivid discussion phase, the committee decided to accept 11 papers (which represents an acceptance rate of about 29 %). The conference program included three invited talks and some presentations of work in progress as well.

The excellent facilities provided by the EasyChair conference management system allowed us to deal with the submissions successfully and handle the preparation of these proceedings in time.

We would like to thank all invited speakers and authors for their contributions, the Program Committee and the external reviewers for their cooperation, and Springer for its very professional publishing work.

July 2016

Pavel Král
Carlos Martín-Vide

Organization

SLSP 2016 was organized by the Department of Computer Science and Engineering and the Department of Cybernetics, University of West Bohemia, and the Research Group on Mathematical Linguistics (GRLMC) of Rovira i Virgili University, Tarragona.

Program Committee

Srinivas Bangalore	Interactions LLC, Murray Hill, USA
Roberto Basili	University of Rome Tor Vergata, Italy
Jean-François Bonastre	University of Avignon, France
Nicoletta Calzolari	National Research Council, Pisa, Italy
Marcello Federico	Bruno Kessler Foundation, Trento, Italy
Guillaume Gravier	IRISA, Rennes, France
Gregory Grefenstette	INRIA, Saclay, France
Udo Hahn	University of Jena, Germany
Thomas Hain	University of Sheffield, UK
Dilek Hakkani-Tür	Microsoft Research, Mountain View, USA
Mark Hasegawa-Johnson	University of Illinois, Urbana, USA
Xiaodong He	Microsoft Research, Redmond, USA
Graeme Hirst	University of Toronto, Canada
Gareth Jones	Dublin City University, Ireland
Tracy Holloway King	A9.com, Palo Alto, USA
Tomi Kinnunen	University of Eastern Finland, Joensuu, Finland
Philipp Koehn	University of Edinburgh, UK
Pavel Král	University of West Bohemia, Pilsen, Czech Republic
Claudia Leacock	McGraw-Hill Education CTB, Monterey, USA
Mark Liberman	University of Pennsylvania, Philadelphia, USA
Qun Liu	Dublin City University, Ireland
Carlos Martín-Vide (Chair)	Rovira i Virgili University, Tarragona, Spain
Alessandro Moschitti	University of Trento, Italy
Preslav Nakov	Qatar Computing Research Institute, Doha, Qatar
John Nerbonne	University of Groningen, The Netherlands
Hermann Ney	RWTH Aachen University, Germany
Vincent Ng	University of Texas, Dallas, USA
Jian-Yun Nie	University of Montréal, Canada
Kemal Oflazer	Carnegie Mellon University – Qatar, Doha, Qatar
Adam Pease	Articulate Software, San Francisco, USA
Massimo Poesio	University of Essex, UK
James Pustejovsky	Brandeis University, Waltham, USA
Manny Rayner	University of Geneva, Switzerland
Paul Rayson	Lancaster University, UK

Douglas A. Reynolds	Massachusetts Institute of Technology, Lexington, USA
Erik Tjong Kim Sang	Meertens Institute, Amsterdam, The Netherlands
Murat Saraçlar	Boğaziçi University, Istanbul, Turkey
Björn W. Schuller	University of Passau, Germany
Richard Sproat	Google, New York, USA
Efstathios Stamatatos	University of the Aegean, Karlovassi, Greece
Yannis Stylianou	Toshiba Research Europe Ltd., Cambridge, UK
Marc Swerts	Tilburg University, The Netherlands
Tomoki Toda	Nagoya University, Japan
Xiaojun Wan	Peking University, Beijing, China
Andy Way	Dublin City University, Ireland
Phil Woodland	University of Cambridge, UK
Junichi Yamagishi	University of Edinburgh, UK
Heiga Zen	Google, Mountain View, USA
Min Zhang	Soochow University, Suzhou, China

External Reviewers

Azad Abad
Daniele Bonadiman
Kazuki Irie
Antonio Uva

Organizing Committee

Tomáš Hercig	Pilsen
Carlos Martín-Vide	Tarragona (Co-chair)
Manuel J. Parra	Granada
Daniel Soutner	Pilsen
Florentina Lilica Voicu	Tarragona
Jan Zelinka	Pilsen (Co-chair)

Identifying Sentiment and Emotion
in Low Resource Languages

(Invited Talk)

Julia Hirschberg and Zixiaofan Yang

Department of Computer Science, Columbia University, New York,
NY 10027, USA
{julia,brenda}@cs.columbia.edu

Abstract. When disaster occurs, online posts in text and video, phone messages, and even newscasts expressing distress, fear, and anger toward the disaster itself or toward those who might address the consequences of the disaster such as local and national governments or foreign aid workers represent an important source of information about where the most urgent issues are occurring and what these issues are. However, these information sources are often difficult to triage, due to their volume and lack of specificity. They represent a special challenge for aid efforts by those who do not speak the language of those who need help especially when bilingual informants are few and when the language of those in distress is one with few computational resources. We are working in a large DARPA effort which is attempting to develop tools and techniques to support the efforts of such aid workers very quickly, by leveraging methods and resources which have already been collected for use with other, High Resource Languages. Our particular goal is to develop methods to identify sentiment and emotion in spoken language for Low Resource Languages.

Our effort to date involves two basic approaches: (1) training classifiers to detect sentiment and emotion in High Resources Languages such as English and Mandarin which have relatively large amounts of data labeled with emotions such as anger, fear, and stress and using these directly of adapted with a small amount of labeled data in the LRL of interest, and (2) employing a sentiment detection system trained on HRL text and adapted to the LRL using a bilingual lexicon to label transcripts of LRL speech. These labels are then used as labels for the aligned speech to use in training a speech classifier for positive/negative sentiment. We will describe experiments using both such approaches, comparing each to training on manually labeled data.

Contents

Invited Talks

Continuous-Space Language Processing: Beyond Word Embeddings

Mari Ostendorf[(✉)]

Electrical Engineering Department, University of Washington, Seattle, USA
ostendor@uw.edu

Abstract. Spoken and written language processing has seen a dramatic shift in recent years to increased use of continuous-space representations of language via neural networks and other distributional methods. In particular, word embeddings are used in many applications. This paper looks at the advantages of the continuous-space approach and limitations of word embeddings, reviewing recent work that attempts to model more of the structure in language. In addition, we discuss how current models characterize the exceptions in language and opportunities for advances by integrating traditional and continuous approaches.

Keywords: Word embeddings · Continuous-space language processing · Compositional language models

1 Introduction

Word embeddings – the projection of word indicators into a low-dimensional continuous space – have become very popular in language processing. Typically, the projections are based on the distributional characteristics of the words, e.g. word co-occurence patterns, and hence they are also known as distributional representations. Working with words in a continuous space has several advantages over the standard discrete representation. Discrete representations lead to data sparsity, and the non-parametric distribution models people typically use for words do not have natural mechanisms for parameter tying. While there are widely used algorithms for learning discrete word classes, these are based on maximizing mutual information estimated with discrete distributions, which gives a highly biased estimate at the tails of the distribution leading to noise in the class assignments. With continuous-space models, there are a variety of techniques for regularization that can be used, and the distributional representation is effectively a soft form of parameter sharing. The distributed representation also provides a natural way of computing word similarity which gives a reasonable match to human judgements even with unsupervised learning. In a discrete space, without distributional information, all words are equally different. Continuous-space representations are also better suited for use in multi-modal applications. Continuous-space language processing has facilitated an explosive growth in work combining images and natural language, both for applications

© Springer International Publishing AG 2016
P. Král and C. Martín-Vide (Eds.): SLSP 2016, LNAI 9918, pp. 3–15, 2016.
DOI: 10.1007/978-3-319-45925-7_1

such as image captioning [18, 33] as well as richer resources for learning embedded representations of language [8]. Together with advances in the use of neural networks in speech recognition, continuous-space language models are also opening new directions for handling open vocabularies in speech recognition [9, 47]. Lastly, there is a growing number of toolkits (e.g. Theano, TensorFlow) that make it easy to get started working in this area.

Despite these important advantages, several drawbacks are often raised to using word embeddings and neural networks more generally. One concern is that neural language processing requires a large amount of training data. Of course, we just argued above that discrete models are more sensitive to data sparsity. A typical strategy for discrete language models is to leverage morphology, but continuous-space models can in fact leverage this information more effectively for low resource languages [19]. Another concern is that representing a word with a single vector is problematic for words with multiple senses. However, Li and Jurafsky [42] show that larger dimensions and more sophisticated models can obviate the need for explicit sense vectors. Yet another concern is that language is compositional and the popular sequential neural network models do not explicitly represent this, but the field is in its infancy and already some compositional models exist [15, 62]. In addition, the currently popular deep neural network structures can be used in a hierarchical fashion, as with character-based word models discussed here or hierarchical sentence-word sequence models [43]. Even with sequential models, analyses show that embeddings can learn meaningful structure.

Perhaps the biggest concern about word embeddings (and their higher level counterparts) is that the models are not very interpretable. However, the distributional representations are arguably more interpretable than discrete representations. While one cannot traceback from a single embedding element to a particular word or set of words (unless using non-negative sparse embeddings [54]), nearest-neighbor examples are often effective for highlighting model differences and failures. Visualizations of embeddings [48] can illustrate what is being learned. Neural networks used as a black box are uninterpretable, but work aiming to link deep networks with generative statistical models holds promise for building more interpretable networks [24]. And some models are more interpretable than other: convolutional neural network filter outputs and attention modeling frameworks provide opportunities for analysis through visualization of weights. In addition, there are opportunities for designing architectures that factor models or otherwise incorporate knowledge of properties of language, which can contribute to interpretability and improve performance. Outlining these opportunities is a primary goal of this paper.

A less discussed problem with continuous-space models is that the very property that makes them good at learning regularities and ignoring noise such as typographical errors makes them less well suited to learning the exceptions or idiosyncracies in human language. These exceptions occur at multiple linguistic levels, e.g. irregular verb conjugations, multi-word expressions, idiomatic expressions, self-corrections, code switching and social conventions. Human language learners are taught to simply memorize the exceptions. Discrete models are well suited to handling such cases. Is there a place for mixed models?

In the remainder of the paper, we overview a variety of approaches to continuous-space representation of language with an emphasis on characterizing structure in language and providing evidence that the models are indeed learning something about language. We first review popular approaches for learning word embeddings in Sect. 2, discussing the success and limitations of the vector space model, and variations that attempt to capture multiple dimensions of language. Next, in Sect. 3, we discuss character-based models for creating word embeddings that provide more compact models and provide open vocabulary coverage. Section 4 looks at methods and applications for sentence-level modeling, particularly those with different representations of context. Finally, Sect. 5 closes with a discussion of a relatively unexplored challenge in this field: characterizing the idiosyncrasies and exceptions of language.

2 Word Embeddings

The idea of characterizing words in terms of their distributional properties has a long history, and vector space models for information retrieval date back to the 70's. Examples of their use in early automatic language processing work includes word sense characterization [60] and multiple choice analogy tests [67]. Work by Bengio and colleagues [4,5] spawned a variety of approaches to language modeling based on neural networks. Later, Collobert and Weston [12,13] proposed a unified architecture for multiple natural language processing tasks that leverage a single neural network bottleneck stage, i.e. that share word embeddings. In [50], Mikolov and colleagues demonstrated that word embeddings learned in an *unsupervised* way from a recurrent neural network (RNN) language model could be used with simple vector algebra to find similar words and solve analogy problems. Since then, several different unsupervised methods for producing word embeddings have been proposed. Two popular methods are based on word2vec [49] and GloVe [55]. In spite of the trend toward deep neural networks, these two very successful models are in fact shallow: a one-layer neural network and a logbilinear model, respectively. One possible explanation for their effectiveness is that the relative simplicity of the model allows them to be trained on very large amounts of data. In addition, it turns out that simple models are compatible with vector space similarity measures.

In computing word similarity with word embeddings, typically either a cosine distance ($\cos(x,y) = x^t y/(||x|| \cdot ||y||)$) or Euclidean distance ($d(x,y) = ||x - y||$) are used. (Note that for unit norm vectors, $||x|| = ||y|| = 1$, $\arg\max_x \cos(x,y) = \arg\min_x d(x,y)$.) Such choices seem reasonable for a continuous space, but other distances could be used. If x was a probability distribution, Euclidean distance would not necessarily be a good choice. To better motivate the choice, consider a particular approach for generating embeddings, the logbilinear model. Let x (and y) be the one-hot indicator of a word and $\tilde{x}$ (and $\tilde{y}$) be its embedding (projection to a lower dimensional space). Similarly, w indicates word context and $\tilde{w}$ its projection. In the logbilinear model,

$$\log p(w,x) = K + x^t A w = K + x^t U^t V w = K + \tilde{x}^t \tilde{w}.$$

(In a discrete model, A could be full rank. The projections characterize a lower rank that translate into shared parameters across different words [27].) Training this model to maximize likelihood corresponds to training it to maximize the inner product of two word embeddings when they co-occur frequently. Define two words x and y to be similar when they have similar distributional properties, i.e. $p(w, x)$ is close to $p(w, y)$ for all w. This corresponds to a log probability difference: for the log bilinear model, $(\tilde{x} - \tilde{y})^t w$ should be close to 0, in which case it makes sense to minimize the Euclidean distance between the embeddings. More formally, using the minimum Kullback-Leibler (KL) distance as a criterion for closeness of distributions, the logbilinear model results in the criterion

$$\arg\min_y D(p(w|y)\|p(w|x)) = \arg\min_y E_{W|Y}[\log p(w|y) - \log p(w|x)]$$
$$= \arg\min_y E[\tilde{w}|y]^t (\tilde{y} - \tilde{x}) + K_y.$$

Thus, to minimize the KL distance, Euclidean distance is not exactly the right criterion, but it is a reasonable approximation. Since the logbilinear model is essentially a simple, shallow neural network, it is reasonable to assume that this criterion would extend to other shallow embeddings. This representation provides a sort of soft clustering alternative to discrete word clustering [7] for reducing the number of parameters in the model, and the continuous space approach tends to be more robust.

The analogy problem involves finding b such that x is to y as a is to b. The vector space model estimates $\hat{b} = y - x + b$ and finds b according to the maximum cosine distance $\cos(b, \hat{b}) = b^t \hat{b} / |b||\hat{b}|$, which is equivalent to the minimum Euclidean distance when the original vectors have unit norm. In [39], Levy and Goldberg point out that for the case of unit norm vectors,

$$\arg\max_b \cos(b, y - x + a) = \arg\max_b (\cos(b, y) - \cos(b, x) + \cos(b, a)).$$

Thus, maximizing the similarity to the estimated vector is equivalent to choosing word b such that its distributional properties are similar to both words y and a, but dissimilar to x. This function is not justified with the log bilinear model and a minimum distribution distance criterion, consistent with the finding that a modification of the criterion gave better results [39].

A limitation of these models is that they are learning functional similarity of words, so words that can be used in similar contexts but have very different polarities can have an undesirably high similarity (e.g. "pretty," "ugly"). Various directions have been proposed for improving embeddings including, for example, multilingual learning with deep canonical correlation analysis (CCA) [46] and leveraging statistics associated with common word patterns [61]. What these approaches do not capture is domain-specific effects, which can be substantial. For example, the word "sick" could mean ill, awesome, or in bad taste, among other things. For that reason, domain-specific embeddings can give better results than general embeddings when sufficient training data is available. Various toolkits are available; with sufficient tuning of hyperparameters, they can give similar results [40].

Beyond the need for large amounts of training data, learning word embeddings from scratch is unappealing in that, intuitively, much of language should generalize across domains. In [12], shared aspects of language are captured via multi-task training, where the final layers of the neural networks are trained on different tasks and possibly different data, but the lower levels are updated based on all tasks and data. With a simpler model, e.g. a logbilinear model, it is possible to factor the parameters according to labeled characteristics of the data (domain, time, author/speaker) that allow more flexible sharing of parameters across different subsets of data and can be easily jointly trained [16,27,70]. This is a form of capturing structure in language that represents a promising direction for new models.

3   Compositional Character Models

A limitation of word embeddings (as well as discrete representations of words) is the inability to handle words that were unseen in training, i.e. out-of-vocabulary (OOV) words. Because of the Zipfian nature of language, encountering new words is likely, even when sufficient training data is available to use very large vocabularies. Further, use of word embeddings with very large vocabularies typically has a high memory requirement. OOV words pose a particular challenge for languages with a rich morphology and/or minimal online text resources.

One strategy that has been used to address the problem of OOV words and limited training data is to augment the one-hot word input representation with morphological features. (Simply replacing words with morphological features is generally less effective.) Much of the work has been applied to language modeling, including early work with a simple feedforward network [1] and more recently with a deep neural network [53], exponential models [6,19,28], and recurrent neural networks [19,68]. Other techniques have been used to learn embeddings for word similarity tasks by including morphological features, including a recursive neural network [45] and a variant of the continuous bag of words model [56].

All of these approaches rely on the availability of either a morphological analysis tool or a morphologically annotated dictionary for closed vocabulary scenarios. Most rely on Morfessor [14], which is an unsupervised technique for learning a morphological lexicon that has been shown to be very effective for several applications and a number of languages. However, the resulting lexicon does not cover word stems that are unseen in training, and it is less well suited to nonconcatenative morphology. The fact that work on unsupervised learning of word embeddings has been fairly successful raises the question of whether it might be possible to learn morphological structure of words implicitly by characterizing the sequence of characters that comprise a word. This idea and the desire to more efficiently handle larger vocabularies has led to recent work on learning word embeddings via character embeddings.

There are essentially two main compositional models that have been proposed for building word embeddings from character embeddings: recursive neural networks (RNNs) and convolutional neural networks. (CNNs) In both cases, the word embeddings form the input to a word-level RNN, typically a long-short-term-memory (LSTM) network. Work with character-level recurrent neural networks has used bi-directional LSTMs for language modeling and part-of-speech (POS) tagging on five languages [44], dependency parsing on twelve languages [3], and slot filling text analysis in English [29]. The first studies with standard convolutional neural networks addressed problems related to POS tagging for Portuguese and English [59] and named entity recognition for English and Spanish [58]. In [35], Kim et al. use multiple convolutional filters of different lengths and add a "highway network" [65] between the CNN output and the word-level LSTM, which is analogous to the gating function of an LSTM. They obtain improvements in perplexity in six languages compared to both word and word+morph-based embeddings. The same model is applied to the 1B word corpus in English with good results and a substantial decrease in model size [31]. In our own work on language identification, we find good performance is obtained using the CNN architecture proposed by [35]. All those working on multiple languages report that the gains are greatest for morphologically rich languages and infrequent or OOV words. Language model size reductions compared to word-base vocabularies range from roughly a factor of 3 for CNN variants to a factor of 20–30 for LSTM architectures.

Building word embeddings from character embeddings has the advantage of requiring substantially fewer parameters. However, words that appear frequently may not be as effectively represented with the compositional character embeddings. Thus, in some systems [29,59], the word embedding is a concatenation of two sub-vectors: one learned from full words and the other from a compositional character model. In this case, one of the "words" corresponds to the OOV word.

The studies show that the character-based models are effective for natural language processing tasks, but are they learning anything about language? Certainly the ability to handle unseen words is a good indication that they are. However, the more in-depth analyses reported are mixed. For handling OOVs, examples reported are quite encouraging, both for actual new words and spelling variants, e.g. from [35], the nearest neighbor to "computer-aided" is "computer-guided" and to "looooook" is "look." Similarly, [44] reports good results for nonce words: "phding" is similar to in-vocabulary "-ing" words and "Noahshire" is similar to other "-shire" words and city names. Examples from [59] indicate that the models are learning prefixes and suffixes, and [3] finds that words cluster by POS. However, [35] points out that although character combinations learned from the filters tend to cluster in prefix/suffix/hyphenation/other categories, "they did not (in general) correspond to valid morphemes." The highway network leads to more semantically meaningful results, fixing the unsatisfying "while" and "chile" similarity. Thus, it may be that other variants on the architecture will be useful.

The focus of this discussion has been on architectures that create word embeddings from character embeddings, because words are useful in compositional models aiming at phrase or sentence meaning. However, there are applications where it may be possible to bypass words altogether. Unsupervised learning of character embeddings is useful for providing features to a conditional random field for text normalization of tweets [11]. For text classification applications, good results have been obtained by representing a document as a sequence of characters [72] or a bag of character trigrams [26] for text classification applications. Also worth noting: the same ideas can be applied to bytes as well as characters [20] and to mapping sequences of phonemes to words for speech recognition [9,17,47].

4   Sentence Embeddings

Word embeddings are useful for language processing problems where word-level features are important, and they provide a accessible point for analysis of model behavior. However, most NLP applications require processing sentences or documents comprised of sequences of sentences. Because sentences and documents have variable length, one needs to either map the word sequence into a vector or use a sequence model to characterize it for automatic classification. A classic strategy is to characterize text as a bag of words (or a bag of n-grams, or character n-grams). The simple extension in continuous space is to average word vectors. This can work reasonably well at the local level as in the continuous bag-of-words (CBOW) model [49], and there is some work that has successfully used averaging for representing short sentences [23]. However, it is considered to be the wrong use of embeddings for longer spans of text [38,72].

There are a number of approaches that have been proposed for characterizing word sequences, including RNNs [52], hierarchical CNNs [32,34], recursive CNNs [37] and more linguistically motivated alternatives, e.g., recursive neural networks [62,63] and recurrent neural network grammars [15]. Taken together, these different models have been used for a wide variety of language processing problems, from core analysis tasks such as part-of-speech tagging, parsing and sentiment classification to applications such as language understanding, information extraction, and machine translation.

In this work, we focus on RNNs, since most work on augmenting the standard sequence model has been based in this framework, as in the case of the character-based word representations described above. There are a number of RNN variants aimed at dealing with the vanishing gradient problem, e.g. the LSTM [25,66] and versions using a gated recurrent unit [10]. Since these different variants are mostly interchangeable, the term RNN will be used generically to include all such variants.

While there has been a substantial impact from using sentence-level embeddings in many language processing tasks, and experiments with recursive neural networks show that the embedded semantic space does capture similarity of different length paraphrases in a common vector space [63], the single vector model is problematic for long sentences. One step toward addressing this issue is to use bi-dierctional models and concatenate embedding vectors generated in both directions, as for the bi-direction LSTM [22]. For a tree-structured model, upward and downward passes can be used to create two subvectors that are concatenated, as in work on identifying discourse relations [30].

Expanding on this idea, interesting directions for research that characterize sentences with multiple vectors include attention models, factored modeling of context, and segmental models. The neural attention model, as used in machine translation [2] or summarization [57], provides a mechanism for augmenting the sentence-level vector with a context-dependent weighted combination of word models for text generation. A sentence is "encoded" into a single vector using a bi-directional RNN, and the translated version is generated (or "decoded") by this input with a state-dependent context vector that is a weighted sum of word embeddings from the original sentence where the weights are determined using a separate network that learns what words in the encoded sentence to pay attention to given the current decoder state. For the attention model, embeddings for all words in the sentence must be stored in addition to the overall encoding. This can be impractical for long sentences or multi-sentence texts. Context models characterize sentences with multiple sub-vectors corresponding to different factors that contribute to that sentence. For example, [51] learn a context vector using latent Dirichlet analysis to augment an RNN language model. For language generation, neural network context models have characterized conversation history [64], intention [69] and speaker [41] jointly with sentence content. Lastly, segmental models [21,36] identify subvectors associated with an unobserved variable-length segmentation of the sequence.

5   The Future: Handling the Idiosyncracies of Language

This paper has argued that many of the supposed limitations of continuous-space approaches are not really limitations, and shown that the lack of structure in current models is an active area of research with several promising new directions of research. What has received much less attention are the idiosyncracies and exceptions in language. Continuous-space models are essentially low-rank, smoothed representations of language; smoothing tends to minimize exceptions. Of course, exceptions that occur frequently (like irregular verbs or certain misspellings) can be learned very effectively with continuous-space models. Experiments show that idiosyncracies that are systematic, such as typographical exaggerations ("loooooook" for "look") can also be learned with character-based word models.

Other problems have mixed results. Disfluencies, including filled pauses, restarts, repetitions and self-corrections, can be thought of as an idiosyncrasy of spoken language. There is structure in repetitions and to a lesser extent in self-corrections, and there is some systematicity in where disfluencies occur, but they are highly variable. Further, speech requires careful transcription for accurate representation of disfluencies, and there is not a lot of such data available. State-of-the-art performance in disfluency detection has been obtained with bidirectional LSTMs, but only with engineered disfluency pattern match features augmenting the word sequence [71]. Another phenomenon that will likely benefit from feature augmentation is characterization of code-switching. While character-based models from different languages can be combined to handle whole word code-switching, it will be less able to handle the native morphological inflections of non-native words.

The use of factored models allows parameters for general trends to be learned on large amounts of shared data freeing up parameters associated with different factors to characterize idiosyncracies. However, these exceptions by their nature are sparse. One mechanism for accounting for such exceptions is to use a mixed continuous and discrete (or low-rank and sparse) model of language, incorporating L1 regularization for a subset of the parameters. In [27], a sparse set of word-indexed parameters is learned to adjust probabilities for exception words and n-grams, both positively and negatively. The sparse component learns multi-word expressions ("New York" is more frequent than would be expected from their unigram frequencies) as well idiosyncracies of informal speech ("really much" is rarely said, although "really" is similar to "very" and "very much" is a relatively frequent pair).

In summary, the field of language processing has seen dramatic changes and is currently dominated by neural network models. While the black-box use of these methods can be problematic, there are many opportunities for innovation and advances. Several new architectures are being explored that attempt to incorporate more of the hierarchical structure and context dependence of language. At the same time, there are opportunities to integrate the strengths of discrete models and linguistic knowledge with continuous-space approaches to characterize the idiosyncracies of language.

Acknowledgments. I thank my students Hao Cheng, Hao Fang, Ji He, Brian Hutchinson, Aaron Jaech, Yi Luan, and Vicky Zayats for helping me gain insights into continuous space language methods through their many experiments and our paper discussions.

References

1. Alexandrescu, A., Kirchhoff, K.: Factored neural language models. In: Proceedings of the Conference North American Chapter Association for Computational Linguistics: Human Language Technologies (NAACL-HLT) (2006)
2. Bahdanau, D., Cho, K., Bengio, Y.: Neural machine translation by jointly learning to align and translate. In: Proceedings of the International Conference Learning Representations (ICLR) (2015)
3. Ballesteros, M., Dyer, C., Smith, N.: Improved transition-based parsing by modeling characters instead of words with LSTMs. In: Proceedings of the Conference Empirical Methods Natural Language Process (EMNLP), pp. 349–359 (2015)
4. Bengio, Y., Ducharme, R., Vincent, P., Jauvin, C.: A neural probabilistic language model. J. Mach. Learn. Res. **3**, 1137–1155 (2003)
5. Bengio, Y., Ducharme, R., Vincent, P.: A neural probabilistic language model. In: Proceedings of the Conference Neural Information Processing System (NIPS), pp. 932–938 (2001)
6. Botha, J.A., Blunsom, P.: Compositional morphology for word representations and language modelling. In: Proceedings of the International Conference on Machine Learning (ICML) (2014)

7. Brown, P.F., Della Pietra, V.J., de Souza, P.V., Lai, J.C., Mercer, R.L.: Class-based n-gram models of natural language. Comput. Linguist. **18**, 467–479 (1992)
8. Bruni, E., Tran, N., Baroni, M.: Multimodal distributional semantics. J. Artif. Intell. Res. **49**, 1–47 (2014)
9. Chan, W., Jaitly, N., Le, Q., Vinyals, O.: Listen, attend and spell: a neural network for large vocabulary conversational speech recognition. In: Proceedings of the International Conference Acoustic, Speech, and Signal Process (ICASSP), pp. 4960–4964 (2016)
10. Cho, K., van Merriënboer, B., Gulcehre, C., Bahadanau, D., Bougares, F., Schwenk, H., Bengio, Y.: Learning phrase representations using RNN encoder-decoder for statistical machine translation. In: Proceedings of the Conference Empirical Methods Natural Language Process (EMNLP), pp. 1724–1734 (2014)
11. Chrupala, G.: Normalizing tweets with edit scripts and recurrent neural embeddings. In: Proceedings of the Annual Meeting Association for Computational Linguistics (ACL) (2014)
12. Collobert, R., Weston, J.: A unified architecture for natural language processing: deep neural networks with multitask learning. In: Proceedings of the International Conference Machine Learning (ICML), pp. 160–167 (2008)
13. Collobert, R., Weston, J., Bottou, L., Karlen, M., Kavukcuoglu, K., Kuksa, P.: Natural language processing (almost) from scratch. J. Mach. Learn. Res. **12**, 2493–2537 (2011)
14. Creutz, M., Lagus, K.: Inducing the morphological lexicon of a natural language from unannotated text. In: Proceedings International and Interdisciplinary Conference on Adaptive Knowledge Representation and Reasoning (AKRR), June 2005
15. Dyer, C., Kuncoro, A., Ballesteros, M., Smith, N.A.: Recurrent neural network grammars. In: Proceedings of the Conference North American Chapter Association for Computational Linguistics (NAACL) (2015)
16. Eisenstein, J., Ahmed, A., Xing, E.P.: Sparse additive generative models of text. In: Proceedings of the International Conference Machine Learning (ICML) (2011)
17. Eyben, F., Wöllmer, M., Schuller, B., Graves, A.: From speech to letters - using a novel neural network architecture for grapheme based ASR. In: Proceedings of the Automatic Speech Recognition and Understanding Workshop (ASRU), pp. 376–380 (2009)
18. Fang, H., Gupta, S., Iandola, F., Srivastava, R., Deng, L., Dollar, P., Gao, J., He, X., Mitchell, M., Platt, J., Zitnick, L., Zweig, G., Zitnick, L.: From captions to visual concepts and back. In: Proceedings of the Conference Computer Vision and Pattern Recognition (CVPR) (2015)
19. Fang, H., Ostendorf, M., Baumann, P., Pierrehumbert, J.: Exponential language modeling using morphological featues and multi-task learning. IEEE Trans. Audio Speech Lang. Process. **23**(12), 2410–2421 (2015)
20. Gillick, D., Brunk, C., Vinyals, O., Subramanya, A.: Multilingual language processing from bytes. In: Proceedings of the Conference North American Chapter Association for Computational Linguistics (NAACL) (2016)
21. Graves, A., Fernandez, S., Gomez, F., Schmidhuber, J.: Connectionist temporal classification: labeling unsegmented sequence data with recurrent neural networks. In: Proceedings of the International Conference Machine Learning (ICML) (2006)
22. Graves, A., Schmidhuber, J.: Framewise phoneme classification with bidirectional LSTM and other neural network architectures. Neural Netw. **18**(5), 602–610 (2005)
23. He, J., Chen, J., He, X., Gao, J., Li, L., Deng, L., Ostendorf, M.: Deep reinforcement learning with a natural language action space. In: Proceedings of the Annual Meeting Association for Computational Linguistics (ACL) (2016)

24. Hershey, J.R., Roux, J.L., Weninger, F.: Deep unfolding: model-based inspiration of novel deep architectures. arXiv preprint arXiv:1409.2574v4 (2014)

25. Hochreiter, S., Schmidhuber, J.: Long short-term memory. Neural Comput. **9**(8), 1735–1780 (1997)

26. Huang, P.S., He, X., Gao, J., Deng, L., Acero, A., Heck, L.: Learning deep structured semantic models for web search using clickthrough data. In: Proceedings of the ACM International Conference on Information and Knowledge Management (2013)

27. Hutchinson, B., Ostendorf, M., Fazel, M.: A sparse plus low rank maximum entropy language model for limited resource scenarios. IEEE Trans. Audio Speech Lang. Process. **23**(3), 494–504 (2015)

28. Hutchinson, B.: Rank and sparsity in language processing. Ph.D. thesis, University of Washington, August 2013

29. Jaech, A., Heck, L., Ostendorf, M.: Domain adaptation of recurrent neural networks for natural language understanding. In: Proceedings of the International Conference Speech Communication Association (Interspeech) (2016)

30. Ji, Y., Eisenstein, J.: One vector is not enough: entity-augmented distributional semantics for discourse relations. Trans. Assoc. Comput. Linguist. (TACL) **3**, 329–344 (2015)

31. Jozafowicz, R., Vinyals, O., Schuster, M., Shazeer, N., Wu, Y.: Exploring the limits of language modeling. arXiv preprint arXiv:1602.02410 (2015)

32. Kalchbrenner, N., Grefenstette, E., Blunsom, P.: A convolutional neural network for modelling sentences. In: Proceedings of the Annual Meeting Association for Computational Linguistics (ACL) (2014)

33. Karpathy, A., Fei-Fei, L.: Deep visual-semantic alignments for generating image descriptions. In: Proceedings of the Conference Computer Vision and Pattern Recognition (CVPR) (2015)

34. Kim, Y.: Convolutional neural networks for sentence classification. In: Proceedings of the Conference Empirical Methods Natural Language Process (EMNLP) (2014)

35. Kim, Y., Jernite, Y., Sontag, D., Rush, A.: Character-aware neural language models. In: Proceedings of the AAAI, pp. 2741–2749 (2016)

36. Kong, L., Dyer, C., Smith, N.: Segmental neural networks. In: Proceedings of the International Conference Learning Representations (ICLR) (2016)

37. Lai, S., Xu, L., Liu, K., Zhao, J.: Recurrent convolutional neural networks for text classification. In: Proceedings of the AAAI (2015)

38. Lev, G., Klein, B., Wolf, L.: In defense of word embedding for generic text representation. In: International Conference on Applications of Natural Language to Information Systems, pp. 35–50 (2015)

39. Levy, O., Goldberg, Y.: Linguistic regularities in sparse and explicit word representations. In: Proceedings of the Conference Computational Language Learning, pp. 171–180 (2014)

40. Levy, O., Goldberg, Y., Dagan, I.: Improving distributional similarity with lessons learned from word embeddings. In: Proceedings of the Annual Meeting Association for Computational Linguistics (ACL), pp. 211–225 (2015)

41. Li, J., Galley, M., Brockett, C., Gao, J., Dolan, B.: A persona-based neural conversation model. In: Proceedings of the Annual Meeting Association for Computational Linguistics (ACL) (2016)

42. Li, J., Jurafsky, D.: Do multi-sense embeddings improve natural language understanding? In: Proceedings of the Conference North American Chapter Association for Computational Linguistics (NAACL), pp. 1722–1732 (2015)

43. Lin, R., Liu, S., Yang, M., Li, M., Zhou, M., Li, S.: Hierarchical recurrent neural network for document modeling. In: Proceedings of the Conference Empirical Methods Natural Language Processing (EMNLP), pp. 899–907 (2015)

44. Ling, W., Luís, T., Marujo, L., Astudillo, R.F., Amir, S., Dyer, C., Black, A.W., Trancoso, I.: Finding function in form: compositional character models for open vocabulary word representation. In: EMNLP (2015)

45. Long, M.T., Socher, R., Manning, C.: Better word representations for recursive neural networks for morphology. In: Proceedings of the Conference Computational Natural Language Learning (CoNLL) (2013)

46. Lu, A., Wang, W., Bansal, M., Gimpel, K., Livescu, K.: Deep multilingual correlation for improved word embeddings. In: Proceedings of the Conference North American Chapter Association for Computational Linguistics (NAACL), pp. 250–256 (2015)

47. Maas, A., Xie, Z., Jurafsky, D., Ng, A.: Lexicon-free conversational speech recognition with neural networks. In: Proceedings of the Conference North American Chapter Association for Computational Linguistics (NAACL), pp. 345–354 (2015)

48. van der Maaten, L., Hinton, G.: Visualizing data using t-SNE. Mach. Learn. Res. **9**, 2579–2605 (2008)

49. Mikolov, T., Chen, K., Corrado, G., Dean, J.: Efficient estimation of word representations in vector space. In: Proceedings of the International Conference Learning Representations (ICLR) (2013)

50. Mikolov, T., Yih, W., Zweig, G.: Linguistic regularities in continuous space word representations. In: Proceedings of the Conference North American Chapter Association for Computational Linguistics: Human Language Technologies (NAACL-HLT) (2013)

51. Mikolov, T., Zweig, G.: Context dependent recurrent neural network language model. In: Proceedings of the IEEE Spoken Language Technologies Workshop (2012)

52. Mikolov, T., Martin, K., Burget, L., Černocký, J., Khudanpur, S.: Recurrent neural network based language model. In: Proceedings of the International Conference Speech Communication Association (Interspeech) (2010)

53. Mousa, A.E.D., Kuo, H.K.J., Mangu, L., Soltau, H.: Morpheme-based feature-rich language models using deep neural networks for LVCSR of Egyptian Arabic. In: Proceedings of the International Conference Acoustic, Speech, and Signal Process (ICASSP), pp. 8435–8439 (2013)

54. Murphy, B., Talukdar, P., Mitchell, T.: Learning effective and interpretable semantic models using non-negative sparse embedding. In: Proceedings of the International Conference Computational Linguistics (COLING) (2012)

55. Pennington, J., Socher, R., Manning, C.: GloVe: global vectors for word representation. In: Proceedings of the Conference Empirical Methods Natural Language Process (EMNLP) (2014)

56. Qui, S., Cui, Q., Bian, J., Gao, B., Liu, T.Y.: Co-learning of word representations and morpheme representations. In: Proceedings of the International Conference Computational Linguistics (COLING) (2014)

57. Rush, A., Chopra, S., Weston, J.: A neural attention model for sentence summarization. In: Proceedings of the International Conference Empirical Methods Natural Language Process (EMNLP), pp. 379–389 (2015)

58. dos Santos, C., Guimarães, V.: Boosting named entity recognition with neural character embeddings. In: Proceedings of the ACL Named Entities Workshop, pp. 25–33 (2015)

59. dos Santos, C., Zadrozny, B.: Learning character-level representations for part-of-speech tagging. In: Proceedings of the International Conference Machine Learning (ICML) (2015)
60. Schutze, H.: Automatic word sense discrimination. Comput. Linguist. **24**(1), 97–123 (1998)
61. Schwartz, R., Reichart, R., Rappoport, A.: Symmetric pattern-based word embeddings for improved word similarity prediction. In: Proceedings of the Conference Computational Language Learning, pp. 258–267 (2015)
62. Socher, R., Bauer, J., Manning, C.: Parsing with compositional vectors. In: Proceedings of the Annual Meeting Association for Computational Linguistics (ACL) (2013)
63. Socher, R., Lin, C., Ng, A., Manning, C.: Parsing natural scenes and natural language with recursive neural networks. In: Proceedings of the International Conference Machine Learning (ICML) (2011)
64. Sordoni, A., Galley, M., Auli, M., Brockett, C., Ji, Y., Mitchell, M., Nie, J.Y., Gao, J., Dolan, B.: A neural network approach to context-sensitive generation of conversational responses. In: Proceedings of the Conference North American Chapter Association for Computational Linguistics (NAACL) (2015)
65. Srivastava, R., Greff, K., Schmidhuber, J.: Training very deep networks. In: Proceedings of the Conference Neural Information Processing System (NIPS) (2015)
66. Sundermeyer, M., Schlüter, R., Ney, H.: LSTM neural networks for language modeling. In: Proceedings of the Interspeech (2012)
67. Turney, P.: Similarity of semantic relations. Comput. Linguist. **32**(3), 379–416 (2006)
68. Wu, Y., Lu, X., Yamamoto, H., Matsuda, S., Hori, C., Kashioka, H.: Factored language model based on recurrent neural network. In: Proceedings of the International Conference Computational Linguistics (COLING) (2012)
69. Yao, K., Zweig, G., Peng, B.: Intention with attention for a neural network conversation model. arXiv preprint arXiv:1510.08565v3 (2015)
70. Yogatama, D., Wang, C., Routledge, B., Smith, N., Xing, E.: Dynamic language models for streaming text. Trans. Assoc. Comput. Linguist. (TACL) **2**, 181–192 (2014)
71. Zayats, V., Ostendorf, M., Hajishirzi, H.: Disfluency detection using a bidirectional LSTM. In: Proceedings of the International Conference Speech Communication Association (Interspeech) (2016)
72. Zhang, X., Zhao, J., LeCun, Y.: Character-level convolutional networks for text classification. In: Proceedings of the Conference Neural Information Processing System (NIPS), pp. 1–9 (2015)

Language

Testing the Robustness of Laws of Polysemy and Brevity Versus Frequency

Antoni Hernández-Fernández[2]([✉]), Bernardino Casas[1],
Ramon Ferrer-i-Cancho[1], and Jaume Baixeries[1]

[1] Complexity and Quantitative Linguistics Lab,
Laboratory for Relational Algorithmics, Complexity and Learning (LARCA),
Departament de Ciències de la Computació, Universitat Politècnica de Catalunya,
Barcelona, Catalonia, Spain
{bcasas,rferrericancho,jbaixer}@cs.upc.edu
[2] Complexity and Quantitative Linguistics Lab,
Laboratory for Relational Algorithmics, Complexity and Learning (LARCA),
Institut de Ciències de l'Educació, Universitat Politècnica de Catalunya,
Barcelona, Catalonia, Spain
antonio.hernandez@upc.edu

Abstract. The pioneering research of G.K. Zipf on the relationship between word frequency and other word features led to the formulation of various linguistic laws. Here we focus on a couple of them: the meaning-frequency law, i.e. the tendency of more frequent words to be more polysemous, and the law of abbreviation, i.e. the tendency of more frequent words to be shorter. Here we evaluate the robustness of these laws in contexts where they have not been explored yet to our knowledge. The recovery of the laws again in new conditions provides support for the hypothesis that they originate from abstract mechanisms.

Keywords: Zipf's law · Polysemy · Brevity · Word frequency

1 Introduction

The linguist George Kingsley Zipf (1902–1950) is known for his investigations on statistical laws of language [20,21]. Perhaps the most popular one is **Zipf's law for word frequencies** [20], that states that the frequency of the i-th most frequent word in a text follows approximately

$$f \propto i^{-\alpha} \tag{1}$$

where f is the frequency of that word, i their rank or order and α is a constant ($\alpha \approx 1$). Zipf's law for word frequencies can be explained by information theoretic models of communication and is a robust pattern of language that presents invariance with text length [9] but dependency with respect to the linguistic units considered [5]. The focus of the current paper are a couple of linguistic laws that are perhaps less popular:

© Springer International Publishing AG 2016
P. Král and C. Martín-Vide (Eds.): SLSP 2016, LNAI 9918, pp. 19–29, 2016.
DOI: 10.1007/978-3-319-45925-7_2

- **Meaning-frequency law** [19], the tendency of more frequent words to be more polysemous.
- **Zipf's law of abbreviation** [20], the tendency of more frequent words to be shorter or smaller.

These laws are examples of laws that where the predictor is word frequency and the response is another word feature. These laws are regarded as universal although the only evidence of their universality is that they hold in every language or condition where they have been tested. Because of their generality, these laws have triggered modelling efforts that attempt to explain their origin and support their presumable universality with the help of abstract mechanisms or linguistic principles, e.g., [8]. Therefore, investigating the conditions under which these laws hold is crucial.

In this paper we contribute to the exploration of different definitions of word frequency and word polysemy to test the robustness of these linguistic laws in English (taking into account in our analysis only content words (nouns, verbs, adjectives and adverbs)). Concerning word frequency, in this preliminary study, we consider three major sources of estimation: the CELEX lexical database [3], the CHILDES database [16] and the SemCor corpus[1]. The estimates from the CHILDES database are divided into four types depending on the kind of speakers: children, mothers, fathers and investigators. Concerning polysemy, we consider two related measures: the number of synsets of a word according to WordNet [6], that we refer to as WordNet polysemy, and the number of synsets of WordNet that have appeared in the SemCor corpus, that we refer to as SemCor polysemy. These two measures of polysemy allow one to capture two extremes: the full potential number of synsets of a word (WordNet polysemy) and the actual number of synsets that are used (SemCor polysemy), being the latter a more conservative measure of word polysemy motivated by the fact that, in many cases, the number of synsets of a word overestimates the number of synsets that are known to an average speaker of English. In this study, we assume the polysemy measure provided by Wordnet, although we are aware of the inherent difficulties of borrowing this conceptual framework (see [12,15]). Concerning word length we simply consider orthographic length. Therefore, the SemCor corpus contains SemCor polysemy and SemCor frequency, as well as the length of its lemmas, and the CHILDES database contains CHILDES frequency, the length of its lemmas, and has been enriched with CELEX frequency, WordNet polysemy, and SemCor polysemy. The conditions above lead to $1 + 2 \times 2 = 5$ major ways of investigating the meaning-frequency law and to $1 + 2 = 3$ ways of investigating the law of abbreviation (see details in Sect. 3). The choice made in this preliminary study should not be considered a limitation, since we plan to extend the range of data sources and measures in future studies (we explain these possibilities in Sect. 5).

In this paper, we investigate these laws qualitatively using measures of correlation between two variables. Thus, the law of abbreviation is defined

---

[1] http://multisemcor.fbk.eu/semcor.php.

as a significant negative correlation between the frequency of a word and its length. The meaning-frequency law is defined as a significant positive correlation between the frequency of a word and its number of synsets, a proxy for the number of meanings of a word. We adopt these correlational definitions to remain agnostic about the actual functional dependency between the variable, which is currently under revision for various statistical laws of language [1]. We will show that a significant correlation of the right sign is found in all the combinations of conditions mentioned above, providing support for the hypothesis that these laws originate from abstract mechanisms.

2   Materials

In this section we describe the different corpora and tools that have been used in this paper. We first describe the WordNet database and CELEX corpus, which have been used to compute polysemy and frequency measures. Then, we describe the two different corpora that are analyzed in this paper: SemCor and CHILDES.

2.1   Lexical Database WordNet

The WordNet database [6] can be seen as a set of senses (also called synsets) and relationships among them, where a synset is the representation of an abstract meaning and is defined as a set of words having (at least) the meaning that the synset stands for. Apart from this pair of sets, a relationship between both is also contained. Each pair word-synset is also related to a syntactical category. For instance, the pair *book* and the synset *a written work or composition that has been published* are related to the category *noun*, whereas the pair *book* and synset *to arrange for and reserve (something for someone else) in advance* are related to the category *verb*. WordNet has 155,287 lemmas and 117,659 synsets and contains only four main syntactic categories: nouns, verbs, adjectives and adverbs.

2.2   CELEX Corpus

CELEX [3] is a text corpora in Dutch, English and German, but in this paper we only use the information in English. For each language, CELEX contains detailed information on orthography, phonology, morphology, syntax (word class) and word frequency, based on resent and representative text corpora.

2.3   SemCor Corpus

SemCor is a corpus created at Princeton University composed of 352 texts which are a subset of the English Brown Corpus. All words in the corpus have been sintactically tagged using Brill's part of speech tagger. The semantical tagging has been done manually, mapping all nouns, verbs, adjectives and adverbs, to their corresponding synsets in the WordNet database.

SemCor contains 676, 546 tokens, 234, 136 of which are tagged. In this article we only analyze content words (nouns, verbs, adjectives and adverbs), thus it yields 23, 341 different tagged lemmas that represent only content words.

We use the SemCor corpus to obtain a new measure of polysemy.

SemCor corpus is freely available for download at http://web.eecs.umich.edu/~mihalcea/downloads.html#semcor (accessed 10 August 2016).

2.4 CHILDES Database

The CHILDES database [16] is a set of corpora of transcripts of conversations between children and adults. The corpora included in this database are in different languages, and contains conversations when the children were between 12 and 65 months old, approximately. In this paper we have studied the conversations of 60 children in English (detailed information on these conversations can be found in [4]).

We analyze syntactically every conversation of the selected corpora of CHILDES using Treetagger in order to obtain the lemma and part-of-speech for every word. We have for each word from CHILDES said for each role: lemma, part-of-speech, frequency (number of times that this word is said by this role), number of synsets (according to both SemCor or WordNet), and the word length. We only have taken into account content words (nouns, verbs, adjectives and adverbs). Figure 1 shows the amount of different lemmas obtained from the selected corpora of CHILDES and the amount of analyzed lemmas in this paper for each category. The amount of analyzed lemmas from this corpus is smaller than the total number of lemmas because we have only analyzed those lemmas that are also present in the SemCor corpus.

Role	Tokens	# Lemmas	# Analyzed Lemmas
Child	1, 358, 219	7, 835	4, 675
Mother	2, 269, 801	11, 583	6, 962
Father	313, 593	6, 135	4, 203
Investigator	182, 402	3, 659	2, 775

Fig. 1. Number of tokens, lemmas and analyzed lemmas obtained from CHILDES conversations for each role.

3   Methods

In this paper we compute the relationship between three variables that are related to every lemma: length, frequency and polysemy.

3.1   Length

For the length, we compute the number of letters of the lexical item. Blanks, separation characters and the like have not been taken into consideration.

3.2 Frequency

We have calculated the frequency from three different sources:

- **SemCor frequency**. We use the frequency of each pair *lemma, syntactic category* that is present in the SemCor dataset.
- **CELEX frequency**. We use the frequency of each pair *lemma, syntactic category* that is present in the CELEX lexicon.
- **CHILDES frequency**. For each pair *lemma, syntactic category* that appears in the CHILDES database, we compute its frequency according to each role: child, mother, father, investigator. For example, for the pair *book, noun* we count four different frequencies: the number of times that this pair appears uttered by a child, a mother, a father and an investigator, respectively.

SemCor frequency can only be analyzed in the SemCor corpus, whereas CELEX and CHILDES frequencies are only analyzed in the CHILDES corpora.

3.3 Polysemy

We have calculated the polysemy from two different sources:

- **SemCor polysemy**. For each pair *lemma, syntactic category* we compute the number of different synsets with which this pair has been tagged in the SemCor corpus. This measure is analyzed in the SemCor corpus and in the CHILDES corpus.
- **WordNet polysemy**. For each pair *lemma, syntactic category* we consider the number of synsets according to the WordNet database. This measure is only analyzed in the CHILDES corpus.

We are aware that using a SemCor polysemy measure in the CHILDES corpus or using Wordnet polysemy in both SemCor and CHILDES corpora induces a bias. In the former case, because we are assuming that the same meanings that are used in written text are also used in spoken language. In the latter case, because we are using all possible meanings of a word. An alternative would have been to tag manually all corpora (which is currently an unavailable option) or use an automatic tagger. But also in this case, the possibility of biases or errors would be present. We have performed these combinations for the sake of completeness, and also assuming their limitations.

3.4 Statistical Methods

To compute the relationship between (1) frequency and polysemy and (2) frequency and length. Since frequency and polysemy have more than one source, we have computed all available combinations. In this paper, for the SemCor corpus we analyze the relationship between:

1. SemCor frequency and SemCor polysemy.
2. SemCor frequency and lemma length in the SemCor corpus.

As for the CHILDES corpora, the availability of different sources for frequency and polysemy yields the following combinations:

1. CELEX frequency and SemCor polysemy.
2. CELEX frequency and WordNet polysemy.
3. CHILDES frequency and SemCor polysemy.
4. CHILDES frequency and WordNet polysemy.
5. CHILDES frequency and lemma length in the CHILDES corpus.
6. CELEX frequency and lemma length in the CHILDES corpus.

For each combination of two variables, we compute:

1. **Correlation test**. Pearson, Spearman and Kendall correlation tests, using the `cor.test` standardized R function.
2. **Plot**, in logarithmic scale, that also shows the density of points.
3. **Nonparametric regression**, using the `locpoly` standarized R function, which has been overlapped in the previous plot.

We remark that the analysis for the CHILDES corpora has been segmented by role.

4   Results

We analyze the relationship between (1) frequency and polysemy and (2) frequency and length separately in two different corpora (SemCor and CHILDES).

In both corpora, we have computed a correlation test and a nonparametric regression, which has been plotted alongside with the values of the two variables that are analyzed.

For the SemCor corpus, we have analyzed the relationship between the SemCor frequency and the SemCor polysemy and the relationship between the SemCor frequency and the length of lemmata.

As for the CHILDES corpora, we have analyzed the relationship between two different measures of frequency (CHILDES and CELEX) versus two different measures of polysemy (WordNet and SemCor) and also, the relationship between two different measures of frequency (CHILDES and CELEX) and the length of lemmas. The analysis of individual roles (child, mother, father and investigator) does not show any significant difference between them. In **all** cases we have that:

1. The value of the correlation is *positive* for the relationships frequency-polysemy (see Fig. 2), and *negative* for the relationships frequency-length (see Fig. 4) for all types of correlation: Pearson, Spearman and Kendall. We remark that the p-value is *near zero* in all cases. This is, all correlations are significant.
2. The nonparametric regression function draws a line with a *positive* slope for the frequency-polysemy relationship (see Fig. 3), and *negative* slope for the frequency-length relationship (see Fig. 5). When we say that *it draws a line*, we mean that this function is a quasi-line in the central area of the graph, where most of the points are located. This tendency is not maintained at the extreme parts of graph, where the density of points is significantly lower.

Corpus	ρ	ρ_S	τ_K	Corpus length
SemCor frequency versus SemCor polysemy				
SemCor	0.209	0.627	0.555	23341
CHILDES frequency versus CELEX polysemy				
CHILDES (children)	0.084	0.249	0.177	4675
CHILDES (mothers)	0.081	0.281	0.202	6962
CHILDES (fathers)	0.084	0.279	0.202	4203
CHILDES (investigators)	0.062	0.211	0.153	2775
CELEX frequency versus WordNet polysemy				
CHILDES (children)	0.073	0.353	0.249	4406
CHILDES (mothers)	0.085	0.366	0.261	6577
CHILDES (fathers)	0.089	0.373	0.264	3989
CHILDES (investigators)	0.075	0.341	0.24	2654
CHILDES frequency versus SemCor polysemy				
CHILDES (children)	0.211	0.230	0.178	4675
CHILDES (mothers)	0.186	0.252	0.197	6962
CHILDES (fathers)	0.201	0.256	0.200	4203
CHILDES (investigators)	0.189	0.219	0.171	2775
CELEX frequency versus SemCor polysemy				
CHILDES (children)	0.201	0.607	0.477	4406
CHILDES (mothers)	0.197	0.602	0.474	6577
CHILDES (fathers)	0.226	0.595	0.463	3989
CHILDES (investigators)	0.228	0.585	0.451	2654

Fig. 2. Summary of the analysis of the correlation between the frequency and polysemy of each lemma. Three statistics are considered: the sample Pearson correlation coefficient (ρ), the sample Spearman correlation coefficient (ρ_S) and the sample Kendall correlation tau (τ_K). All correlation tests indicates a significant negative correlation with p-values under 10^{16}.

5   Discussion and Future Work

In this paper, we have reviewed two linguistic laws that we owe to Zipf's [19,20] and that have probably been shadowed by the best-known Zipf's law for word frequencies [20]. Our analysis of the correlation between brevity (measured in number of characters) and polysemy (number of synsets) versus lemma frequency was conducted with three tests with varying assumptions and robustness. Pearson's method supposes input vectors approximately normally distributed while Spearman's is a non-parametric test that does require vectors being approximately normally distributed [2]. Kendall's tau is more robust to extreme observations and to non-linearity compared with the standard Pearson product-moment correlation [17]. Our analysis confirm that a positive correlation between the frequency of the lemmas and the number of synsets (consistent with the meaning-frequency law) and a negative correlation between the length of the lemmas and their frequency (consistent with the law of abbreviation) arises under different

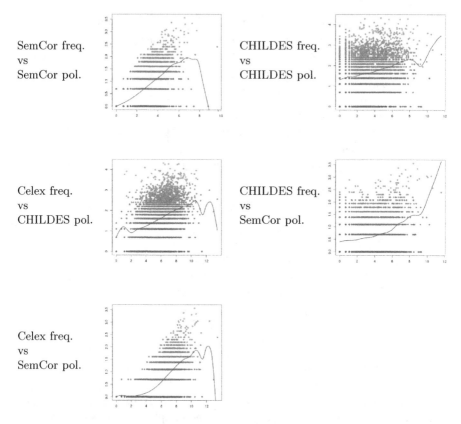

SemCor freq.
vs
SemCor pol.

CHILDES freq.
vs
CHILDES pol.

Celex freq.
vs
CHILDES pol.

CHILDES freq.
vs
SemCor pol.

Celex freq.
vs
SemCor pol.

Fig. 3. Graphics of the relation between frequency (x-axis) and polysemy (y-axis), both in logarithmic scale. The color indicates the density of points: dark green is the highest possible density. The blue line is the nonparametric regression performed over the logarithmic values of frequency and polysemy. We show only the graphs for children.

definitions of the variables. Interestingly, we have not found any remarkable qualitative difference in the analysis of correlations for the different speakers (roles) in the Childes database, suggesting that both child speech and the child-directed-speech (the so-called *motherese*) seem to show the same general statistical biases in the use of more frequent words (that tend to be shorter and more polysemous). With this regard, our results agree with Zipf's pioneering discoveries, independently from the corpora analyzed and independently from the source used to measure the linguistic variables.

Our work offers many possibilities for future research:

First, the analysis of more extensive databases, e.g., Wikipedia in the case of word-length versus frequency.

Second, the use of more fine-grained statistical techniques that allow: (1) to unveil differences between sources or between kinds of speakers, (2) to verify that the tendencies that are shown in this preliminary study are correct,

Corpus	ρ	ρ_S	τ_K	Corpus length
SemCor frequency versus lemma length				
SemCor	−0.062	−0.301	−0.229	23341
CHILDES frequency versus lemma length				
CHILDES (children)	−0.099	−0.324	−0.24	4675
CHILDES (mothers)	−0.076	−0.373	−0.278	6962
CHILDES (fathers)	−0.092	−0.366	−0.277	4203
CHILDES (investigators)	−0.096	−0.318	−0.242	2775
CELEX frequency versus lemma length				
CHILDES (children)	−0.091	−0.132	−0.095	4406
CHILDES (mothers)	−0.084	−0.124	−0.089	6577
CHILDES (fathers)	−0.087	−0.142	−0.102	3989
CHILDES (investigators)	−0.099	−0.172	−0.126	2654

Fig. 4. Summary of the analysis of the correlation between the frequency and the lemma length. Three statistics are considered: the sample Pearson correlation coefficient (ρ), the sample Spearman correlation coefficient (ρ_S) and the sample Kendall correlation tau (τ_K). All correlation tests indicates a significant negative correlation with p-values under 10^{16}.

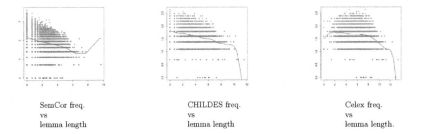

SemCor freq.
vs
lemma length

CHILDES freq.
vs
lemma length

Celex freq.
vs
lemma length.

Fig. 5. Graphics of the relation between frequency (x-axis) and lemma length (y-axis), both in logarithmic scale. The color indicates the density of points: dark green is the highest possible density. The blue line is the nonparametric regression performed over the logarithmic values of frequency and lemma length. We show here only the graphs for children.

and (3) to explain the variations that are displayed in the graphics and to characterize the words that are in the part of the graphics in which our hypotheses hold.

Third, considering different definitions of the same variables. For instance, a limitation of our study is the fact that we define word length using graphemes. An accurate measurement of brevity would require detailed acoustical information that is missing in raw written transcripts [10] or using more sophisticated methods of computation, for instance, to calculate number of phonemes and syllables according to [1]. However, the relationship between the duration of phonemes and graphemes is well-known and in general longer words has longer durations: grapheme-to-phoneme conversion is still a hot topic of research, due to the ambiguity of graphemes with respect to their pronunciation that today supposes a difficulty in speech technologies [18]. In order to improve the frequency measure, we would consider the use of alternative databases, e.g., the frequency of English words in Wikipedia [11].

Forth, our work can be extended including other linguistic variables such as homophony, i.e. words with different origin (and *a priori* different meaning) that have converged to the same phonological form. Actually, Jespersen (1929) suggested a connection between brevity of words and homophony [13], confirmed by Ke (2006) more recently [14] and reviewed by Fenk-Oczlon and Fenk (2010) that outline the *"strong association between shortness of words, token frequency and homophony"* [7].

In fact, the study of different types of polysemy and its multifaceted implications in linguistic networks is descent as future work, as well as the direct study of human voice, because every linguistic phenomenon or candidate for a language law, could be camouflaged or diluted in our transcripts of oral corpus by writing technology, a technology that has been very useful during the last five thousand years, but that prevents us from being close to the acoustic phenomenon of language [10].

Acknowledgments. The authors thank Pedro Delicado and the reviewers for their helpful comments. This research work has been supported by the SGR2014-890 (MACDA) project of the Generalitat de Catalunya, and MINECO project APCOM (TIN2014-57226-P) from Ministerio de Economía y Competitividad, Spanish Government.

References

1. Altmann, E.G., Gerlach, M.: Statistical laws in linguistics. In: Degli Esposti, M., Altmann, E.G., Pachet, F. (eds.) Creativity and Universality in Language. Lecture Notes in Morphogenesis, pp. 7–26. Springer International Publishing, Cham (2016). http://dx.doi.org/10.1007/978-3-319-24403-7_2
2. Baayen, R.H.: Analyzing Linguistic Data: A Practical Introduction to Statistics Using R. Cambridge University Press, Cambridge (2007)

3. Baayen, R.H., Piepenbrock, R., Gulikers, L.: CELEX2, LDC96L14. Philadelphia: Linguistic Data Consortium (1995). https://catalog.ldc.upenn.edu/LDC96L14. Accessed 10 Apr 2016
4. Baixeries, J., Elvevåg, B., Ferrer-i-Cancho, R.: The evolution of the exponent of Zipf's law in language ontogeny. PLoS ONE **8**(3), e53227 (2013)
5. Corral, A., Boleda, G., Ferrer-i Cancho, R.: Zipf's law for word frequencies: word forms versus lemmas in long texts. PLoS ONE **10**(7), 1–23 (2015)
6. Fellbaum, C.: WordNet: An Electronic Lexical Database. MIT Press, Cambridge (1998)
7. Fenk-Oczlon, G., Fenk, A.: Frequency effects on the emergence of polysemy and homophony. Int. J. Inf. Technol. Knowl. **4**(2), 103–109 (2010)
8. Ferrer-i-Cancho, R., Hernández-Fernández, A., Lusseau, D., Agoramoorthy, G., Hsu, M.J., Semple, S.: Compression as a universal principle of animal behavior. Cogn. Sci. **37**(8), 1565–1578 (2013)
9. Font-Clos, F., Boleda, G., Corral, A.: A scaling law beyond Zipf's law and its relation to Heaps' law. New J. Phys. **15**(9), 093033 (2013). http://stacks.iop.org/1367-2630/15/i=9/a=093033
10. Gonzalez Torre, I., Luque, B., Lacasa, L., Luque, J., Hernandez-Fernandez, A.: Emergence of linguistic laws in human voice (2016, in preparation)
11. Grefenstette, G.: Extracting weighted language lexicons from wikipedia. In: Chair, N.C.C., Choukri, K., Declerck, T., Goggi, S., Grobelnik, M., Maegaard, B., Mariani, J., Mazo, H., Moreno, A., Odijk, J., Piperidis, S. (eds.) Proceedings of the Tenth International Conference on Language Resources and Evaluation (LREC 2016). European Language Resources Association (ELRA), Paris, France, May 2016
12. Ide, N., Wilks, Y.: Making sense about sense. In: Agirre, E., Edmonds, P. (eds.) Word Sense Disambiguation: Algorithms and Applications. Text, Speech and Language Technology, vol. 33, pp. 47–73. Springer, Dordrecht (2006). http://dx.doi.org/10.1007/978-1-4020-4809-8_3
13. Jespersen, O.: Monosyllabism in English. Biennial lecture on English philology / British Academy. H. Milford publisher, London (1929). Reprinted in: Linguistica: Selected Writings of Otto Jespersen, pp. 574–598. George Allen and Unwin LTD, London (2007)
14. Ke, J.: A cross-linguistic quantitative study of homophony. J. Quant. Linguist. **13**, 129–159 (2006)
15. Kilgarriff, A.: Dictionary word sense distinctions: an enquiry into their nature. Comput. Humanit. **26**(5), 365–387 (1992). http://dx.doi.org/10.1007/BF00136981
16. MacWhinney, B.: The CHILDES Project: Tools for Analyzing Talk: The Database, vol. 2, 3rd edn. Lawrence Erlbaum Associates, Mahwah (2000)
17. Newson, R.: Parameters behind nonparametric statistics: Kendall's tau, Somers'D and median differences. Stata J. **2**(1), 45–64 (2002)
18. Razavi, M., Rasipuram, R., Magimai-Doss, M.: Acoustic data-driven grapheme-to-phoneme conversion in the probabilistic lexical modeling framework. Speech Commun. **80**, 1–21 (2016)
19. Zipf, G.K.: The meaning-frequency relationship of words. J. Gen. Psychol. **1945**(33), 251–256 (1945)
20. Zipf, G.K.: Human Behaviour and the Principle of Least Effort. Addison-Wesley, Cambridge (1949)
21. Zipf, G.K.: The Psycho-Biology of Language: An Introduction to Dynamic Psychology. MIT Press, Cambridge (1968). Originally published in 1935 by Houghton Mifflin, Boston, MA, USA

Delexicalized and Minimally Supervised Parsing on Universal Dependencies

David Mareček[(✉)]

Institute of Formal and Applied Linguistics,
Faculty of Mathematics and Physics, Charles University in Prague,
Malostranské náměstí 25, 118 00 Praha, Czech Republic
marecek@ufal.mff.cuni.cz

Abstract. In this paper, we compare delexicalized transfer and minimally supervised parsing techniques on 32 different languages from Universal Dependencies treebank collection. The minimal supervision is in adding handcrafted universal grammatical rules for POS tags. The rules are incorporated into the unsupervised dependency parser in forms of external prior probabilities. We also experiment with learning this probabilities from other treebanks. The average attachment score of our parser is slightly lower then the delexicalized transfer parser, however, it performs better for languages from less resourced language families (non-Indo-European) and is therefore suitable for those, for which the treebanks often do not exist.

Keywords: Universal dependenices · Unsupervised parsing · Minimal supervision

1 Introduction

In the last two decades, many dependency treebanks for various languages have been manually annotated. They differ in word categories (POS tagset), syntactic categories (dependency relations), and structure for individual language phenomena. The CoNLL shared tasks for dependency parsing [2,17] unified the file format, and thus the dependency parsers could easily work with 20 different treebanks. Still, the parsing outputs were not comparable between languages since the annotation styles differed even between closely related languages.

In recent years, there have been a huge effort to normalize dependency annotation styles. The Stanford dependencies [11] were adjusted to be more universal across languages [10]. [12] started to develop Google Universal Treebank, a collection of new treebanks with common annotation style using the Stanford dependencies and Universal tagset [19] consisting of 12 part-of-speech tags. [27] produced a collection of treebanks HamleDT, in which about 30 treebanks were automatically converted to a Prague Dependency Treebank style [5]. Later, they converted all the treebanks also into the Stanford style [21].

The researchers from the previously mentioned projects joined their efforts to create one common standard: Universal Dependencies [18]. They used the

© Springer International Publishing AG 2016
P. Král and C. Martín-Vide (Eds.): SLSP 2016, LNAI 9918, pp. 30–42, 2016.
DOI: 10.1007/978-3-319-45925-7_3

Stanford dependencies [10] with minor changes, extended the Google universal tagset [19] from 12 to 17 part-of-speech tags and used the Interset morphological features [25] from the HamleDT project [26]. In the current version 1.2, Universal Dependencies collection (UD) consists of 37 treebanks of 33 different languages and it is very likely that it will continue growing and become common source and standard for many researchers. Now, it is time to revisit the dependency parsing methods and to investigate their behavior on this new unified style.

The goal of this paper is to apply cross language delexicalized transfer parsers (e.g. [14]) on UD and compare their results with unsupervised and minimally supervised parser. Both the methods are intended for parsing languages, for which no annotated treebank exists and both the methods can profit from UD.

In the area of dependency parsing, the term "unsupervised" is understood as that no annotated treebanks are used for training and when supervised POS tags are used for grammar inference, we can deal with them only as with further unspecified types of word.[1] Therefore, we introduce a minimally supervised parser: We use unsupervised dependency parser operating on supervised POS tags, however, we add external prior probabilities that push the inferred dependency trees in the right way. These external priors can be set manually as handwritten rules or trained on other treebanks, similarly as the transfer parsers. This allows us to compare the parser settings with different degrees of supervision:

1. delexicalized training of supervised parsers
2. minimally supervised parser using some external probabilities learned in supervised way
3. minimally supervised parser using a couple of external probabilities set manually
4. fully unsupervised parser

Ideally, the parser should learn only the language-independent characteristics of dependency trees. However, it is hard to define what such characteristics are. For each particular language, we will show what degree of supervision is the best for parsing. Our hypothesis is that a kind of minimally supervised parser can compete with delexicalized transfer parsers.

2   Related Work

There were many papers dealing with delexicalized parsing. [28] transfer a delexicalized parsing model to Danish and Swedish. [14] present a transfer-parser matrix from/to 9 European languages and introduce also multi-source transfer, where more training treebanks are concatenated to form more universal data. Both papers mention the problem of different annotation styles across treebanks, which complicates the transfer. [20] uses already harmonized treebanks [21] and compare the delexicalized parsing for Prague and Stanford annotation styles.

---

[1] In the fully unsupervised setting, we cannot for example simply push verbs to the roots and nouns to become their dependents. This is already a kind of supervision.

Unsupervised dependency parsing methods made a big progress started by the Dependency Model with Valence [7], which was further improved by many other researchers [1, 6, 23, 24]. Many of these works induce grammar based on the gold POS tags, some of them use unsupervised word classes [8, 22]. However, it seems that the research in this field declines in the recent years, probably because its results are still not able to compete with projection and delexicalized methods. An unsupervised grammar induction was joined with a couple of syntactic rules. e.g. in [15] or [3].

3   Data

In all our experiments, we use the Universal Dependencies treebank collection[2] in its current version 1.2. For languages for which there is more than one treebank, we experiment only with the first one.[3] We also exclude 'Japan-KTC' treebank, since the full data are not available. Finally, we experiment with 32 dependency treebanks, each representing a different language. The treebanks, their language families, and their sizes are listed in Table 1.

Before training the parsers, all the treebanks are delexicalized. We substitute all the forms and lemmas by underscores, which are used for undefined values. The same is done with the morphological features and dependency relations. The only information remained is the universal POS tags and the dependency structure (the parent number for each token). The Universal Dependencies use POS tagset consisting of 17 POS tags listed in Table 2.

In the following experiments, we compare delexicalized transfer parsing methods and minimally-supervised methods on the UD treebanks. All the experiments are conducted as if we parsed a language whose syntax is unknown for us. This means that we do not prefer training on syntactically similar languages, we do not prefer right branching or left branching, and do not add language specific word-order rules like preferring SVO or SOV, adjectives before nouns, prepositions vs. postpositions etc.

4   Delexicalized Parsing

We apply the multi-source transfer of delexicalized parser on the UD treebanks in a similar way as [14]. We use the leave-one-out method: for each language, the delexicalized parser is trained on all other treebanks excluding the one on which the parser is tested. Since all the treebanks share the tagset and annotation style, the training data can be simply concatenated together. To decrease the size of the training data and to reduce the training time, we decided to take only first 10,000 tokens for each language, so the final size of the training data

---

[2] universaldependencices.org.
[3] We exclude 'Ancient Greek-PROIEL', 'Finnish-FTB', 'Japan-KTC', 'Latin-ITT', and 'Latin-PROIEL' treebanks.

Table 1. Languages and their families used in the experiments and sizes of the respective treebanks.

Language		Family	Tokens
ar	Arabic	Semitic	282384
bg	Bulgarian	Slavic	156319
cu	Old Slav	Slavic	57507
cs	Czech	Slavic	1503738
da	Danish	Germanic	100733
de	German	Germanic	293088
el	Greek	Hellenic	59156
en	English	Germanic	254830
es	Spanish	Romance	423346
et	Estonian	Uralic	6461
eu	Basque	isolate	121443
fa	Persian	Iranian	151624
fi	Finnish	Uralic	181022
fr	French	Romance	389764
ga	Irish	Celtic	23686
got	Gothic	Germanic	56128
grc	Old Greek	Hellenic	244993
he	Hebrew	Semitic	115535
hi	Hindi	Indo-Iranian	351704
hr	Croatian	Slavic	87765
hu	Hungarian	Uralic	26538
id	Indonesian	Malayic	121923
it	Italian	Romance	252967
la	Latin	Romance	47303
nl	Dutch	Germanic	200654
no	Norwegian	Germanic	311277
pl	Polish	Slavic	83571
pt	Portuguese	Romance	212545
ro	Romanian	Romance	12094
sl	Slovenian	Slavic	140418
sv	Swedish	Germanic	96819
ta	Tamil	Dravidian	9581

is about 300,000 tokens, which is enough for training delexicalized parser. We use the Malt parser[4] [16], and MST parser [13] with several parameter settings. The results are shown in Table 5.

---

[4] Malt parser in the current version 1.8.1 (http://maltparser.org).

Table 2. List of part-of-speech tags used in universal-dependencies treebanks.

ADJ	Adjective	PART	Particle
ADP	Adposition	PRON	Pronoun
ADV	Adverb	PROPN	Proper noun
AUX	Auxiliary verb	PUNCT	Punctuation
CONJ	Coord. conj	SCONJ	Subord. conj.
DET	Determiner	SYM	Symbol
INTJ	Interjection	VERB	Verb
NOUN	Noun	X	Other
NUM	Numeral		

5   Minimally Supervised Parsing

The goal of this paper is to investigate whether the unsupervised parser with added external prior probabilities reflecting the universal annotation scheme is able to compete with the delexicalized methods described in Sect. 4.

We use the unsupervised dependency parser (UDP) implemented by [9]. The reason for this choice was that it has reasonably good results across many languages [8], the source code is freely available,[5] and because it includes a mechanism how to import external probabilities. The UDP is based on Dependency Model with Valence, a generative model which consists of two sub-models:

- Stop model $p_{stop}(\cdot|t_g, dir)$ represents probability of not generating another dependent in direction dir to a node with POS tag t_g. The direction dir can be left or right. If $p_{stop} = 1$, the node with the tag t_g cannot have any dependent in direction dir. If it is 1 in both directions, the node is a leaf.
- Attach model $p_{attach}(t_d|t_g, dir)$ represents probability that the dependent of the node with POS tag t_g in direction dir is labeled with POS tag t_d.

In other words, the *stop* model generates edges, while the *attach* model generates POS tags for the new nodes. The inference is done using blocked Gibbs sampling [4]. During the inference, the *attach* and the *stop* probabilities can be combined linearly with external prior probabilities p^{ext}:

$$p_{stop}^{final} = (1 - \lambda_{stop}) \cdot p_{stop} + \lambda_{stop} \cdot p_{stop}^{ext},$$

$$p_{attach}^{final} = (1 - \lambda_{atach}) \cdot p_{attach} + \lambda_{attach} \cdot p_{attach}^{ext},$$

where the parameters λ define their weights. In the original paper [9], the external priors p_{stop}^{ext} were computed based on the reducibility principle on a big raw corpora.

---

[5] http://ufal.mff.cuni.cz/udp.

5.1 Manually Assigned Priors

We use the external prior probabilities to define grammatical rules for POS tags based on UD annotation style. The first type of priors describes how likely a node with a particular POS is a leaf. We manually set the p_{stop}^{ext} as listed in Table 3. Even though it is possible to define different left and right p_{stop}^{ext}, we decided to set it equally for both the directions, since it is linguistically more language independent.

Table 3. Manual assignment of *stop* probabilities for individual POS tags.

t_g	p_{stop}^{ext}
ADP, ADV, AUX, CONJ, DET, INTJ, NUM, PART, PRON, PUNCT, SCONJ, SYM	1.0
ADJ	0.9
PROPN	0.7
X	0.5
NOUN	0.3
VERB	0.1

In a similar way, we predefine external priors for p_{attach}^{ext}, describing dependency edges.[6] Preliminary experiments showed that less is more in this type of rules. We ended up only with four rules for attaching punctuation and prepositions, as defined in Table 4.[7] Similarly as for p_{stop}^{ext}, we set them equally for both left and right directions. We set $\lambda_{attach} = 0$ for all other possible types of edges, since the priors are not defined for them.

Table 4. Manual assignment of *attach* probabilities for some types of edges.

t_g	t_d	p_{attach}^{ext}
VERB	PUNCT	1.0
NOUN	PUNCT	0.0
VERB	ADP	0.0
NOUN	ADP	1.0

---

[6] We had to change the original parser code to do this.

[7] Note that for example $p_{attach}^{ext}(PUNC|VERB, dir) = 1$ does not mean that all the dependents of VERB must be PUNC. Since the λ_{attach} is less than one, the value 1 only pushes punctuation to be attached below verbs.

5.2 Automatically Assigned Priors

Instead of setting the external probabilities manually, we can compute them automatically from other treebanks. Such experiments are somewhere in the middle between delexicalized parsers and the minimally supervised parser with some manually added knowledge. They learn some regularities but not as many as the delexicalized parsers do.

Similarly as for delexicalized transfer parser, we compute the probabilities on all treebanks but the one which is currently tested. The probabilities are computed in the following way:

$$p_{stop}^{ext}(\cdot|t_g, dir) = \frac{NC(t_g)}{CC(t_g, dir) + NC(t_g)},$$

where $NC(t_g)$ is count of all nodes labelled with tag t_g across all the training treebanks, $CC(t_g, dir)$ is the total number of children in direction dir of all t_g nodes in the treebanks, and

$$p_{attach}^{ext}(t_d|t_g, dir) = \frac{NE(t_g, t_d, dir)}{NE(t_g, *, dir)},$$

where $NE(t_g, t_d, dir)$ is number of dependency edges where the governing node has the POS tag t_g, and the dependent node t_d and is in direction dir from the governing one.

We introduce two additional experiments: *direction-dependent learned priors* (DDLP) and *direction-independent learned priors* (DILP). The external probabilities for DDLP are computed according to the previously mentioned formulas.

In DILP, the probabilities are independent on the direction parameter dir. $p_{stop}^{ext}(\cdot|t_g)$ and $p_{attach}^{ext}(t_d|t_g)$ obtain the same values for both directions. Such approach is therefore less supervised. We suppose, that it gains worse results form majority of languages, however, it could be better for some of languages with word ordering different from the majority of languages.

6   Results

The results of delexicalized transfer parsers, unsupervised parser and minimally supervised parsers with different degrees of supervision on Universal Dependencies are compared in Table 5. We try several settings of parameters for both Malt parser and MST parser, and show the results of two of them for each one.[8] We run the Unsupervised dependency parser by [9], labeled as *UDP*. For UDP, we report four different settings. The *basic* variant is completely unsupervised parsing without any external prior probabilities. The *+rules* column shows the results of our minimally supervised parser (Sect. 5.1) using the external probabilities defined manually (Tables 3 and 4). Both the λ_{stop} and λ_{attach} parameters are set to 0.5. The *DDLP* and *DILP* variants use automatically learned prior probabilities form other treebanks (Sect. 5.2).

---

[8] The results of different parameter settings for both parser varied only little (at most 2 % difference for all the languages).

Table 5. Unlabeled attachment scores for the parsers across the languages. The best results are in bold. For MST parser, we used the second order features and its projective (*proj*) and non-projective (*non-proj*) variant. For the Malt parser, we used lib-SVM training and stacklazy (*lazy*) and nivreeager (*nivre*) algorithms. Unsupervised dependency parser (*UDP*) was tested without any external priors (*basic*), with manual prior probabilities (*+rules*), and with automatically learned probabilities direction dependent (*DDLP*) and direction independent (*DILP*).

Lang.	MST parser		Malt parser		UDP			
	Proj	nproj	Lazy	Nivre	Basic	+Rules	DDLP	DILP
ar	48.8	51.2	50.2	50.4	42.9	51.7	**55.2**	48.0
bg	**79.0**	78.5	78.1	77.4	52.6	74.6	73.2	66.8
cu	64.8	**66.0**	63.1	62.6	46.8	58.1	64.5	59.7
cs	68.0	**68.4**	66.3	65.8	43.6	60.2	62.8	55.4
da	71.0	71.7	66.9	67.2	40 9	57.7	**89.6**	54.8
de	69.8	**70.0**	65.2	65.4	37.4	60.9	63.5	59.8
el	64.3	**64.9**	63.8	64.1	13.1	63.2	62.3	55.9
en	62.1	**62.4**	58.2	58.3	28.1	54.6	54.5	53.0
es	71.5	**72.2**	68.8	69.0	20.4	63.7	66.3	56.1
et	76.4	75.1	70.9	70.5	26.8	79.2	74.6	**80.3**
eu	50.0	51.2	51.8	50.9	47.1	**53.8**	50.5	52.3
fa	52.8	54.8	54.0	54.0	41.0	54.8	**57.5**	45.0
fi	55.1	**55.8**	50.5	50.4	27.6	48.8	46.6	48.7
fr	74.3	**74.8**	71.5	71.5	36.0	65.8	69.0	57.9
ga	60.7	61.4	61.1	61.3	37.1	60.2	**61.5**	57.3
got	63.6	**64.5**	62.8	62.1	47.3	60.2	62.4	57.4
grc	47.2	48.0	45.8	45.5	41.2	50.6	**51.4**	51.2
he	62.5	**64.0**	63.1	62.7	28.2	62.4	**64.0**	56.5
hi	33.5	34.2	35.5	35.1	42.3	50.9	38.4	**54.0**
hr	69.3	**69.4**	67.3	67.1	24.7	61.5	63.4	54.8
hu	57.4	58.0	54.6	54.2	53.4	57.4	55.4	**62.8**
id	58.5	61.0	59.2	58.6	22.7	48.4	**61.3**	51.6
it	76.4	**77.1**	74.0	73.8	42.3	68.8	71.5	60.1
la	**56.5**	55.9	55.5	55.8	47.0	51.8	52.0	47.1
nl	**60.2**	60.1	56.5	57.3	37.5	51.2	54.9	48.5
no	70.2	**70.4**	67.2	66.9	40.9	58.5	61.4	55.7
pl	75.6	**76.0**	74.7	75.0	63.8	68.0	67.7	64.6
pt	73.9	**74.3**	72.4	71.7	40.1	64.6	69.4	58.2
ro	68.3	**69.3**	68.2	67.7	60.4	57.9	66.3	58.9
sl	72.2	**72.8**	71.2	70.6	48.6	68.6	64.9	56.8
sv	70.2	**70.8**	66.2	66.2	41.5	59.5	61.7	58.7
ta	34.3	36.5	35.5	35.6	52.2	52.9	48.4	**58.4**
Avg	63.1	**63.8**	61.6	61.4	39.9	59.4	60.5	56.5

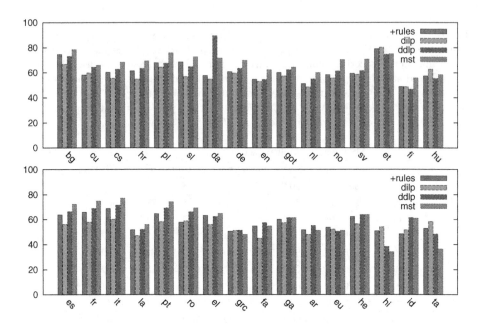

Fig. 1. Comparison of delexicalized parsing methods with different degrees of supervision. UDP with manually set priors (+rules), direction dependent (DDLP) and independent (DILP) learning of priors versus delexicalized transfer of MST parser (mst). Languages are ordered according to their language families: Slavic (bg, cu, cs, hr, pl, sl), Germanic (da, de, en, got, nl, no, sv), Romance (es, fr, it, la, pt, ro), Hellenic (el, grc), Uralic (et, fi, hu), and others (fa, ga, ar, eu, he, hi, id, ta).

7   Discussion

It is evident that the MST parser achieved the best scores. It parsed best 20 out of 32 languages and its non-projective variant reached 63.8 % averaged attachment score. The Malt parser was worse than MST by 2 % in the averaged attachment score.[9] The basic UDP without additional rules performs very poorly, however, with added external prior probabilities, it is competitive with the delexicalized transfer parser methods. 12 out of 32 languages were parsed better by UDP using one variant of the external priors.

With hand-written prior probabilities (*+rules*), the averaged attachment score reached only 59 %, however, it is better than the MST parser on 6 languages: Arabic, Estonian, Basque, Old Greek, Hindi, and Tamil, in two cases by a wide margin. For Persian, the scores are equal.

The averaged attachment score for UDP with direction-independent learned priors (DILP) is even lower (56.5 %), however, it parsed 6 languages better than MST: Estonian, Basque, Old Greek, Hindi, Hungarian, and Tamil.

---

[9] We used the Malt parser with its default feature set. Tuning in this specific delexicalized task would probably bring a bit better results.

Direction dependent learning of priors end up with 60.5 % attachment score and 9 languages better than MST.

Based on these results, we can say that the minimally supervised parser, which takes less information from other annotated treebanks, is more suitable for the more exotic languages, i.e. for languages whose families are less common among the annotated treebanks. Figure 1 shows histograms of attachment scores across languages, now ordered according to the language families. All the Slavic and Romance languages and almost all the Germanic languages[10] are parsed best by the MST parser. Finnish from the three Uralic languages and Greek from the two Hellenic languages are also parsed best by MST. Other 12 languages were better parsed by one of the less supervised methods.

Less-resourced languages, for which the annotated treebanks are missing, may be therefore better parsed by less supervised parsers, especially if they do not belong to the Indo-European language family. The MST transfer parser has probably been over-trained on these Indo-European family languages and is not able to generalize enough to more distant languages. The rules we added to the unsupervised dependency parser (*+rules* experiment) are universal in the direction of dependencies (left/right branching) and cover much more languages.

8   Transfer Parser Comparison Between Different Styles

We compare the best transfer parser results also with the previous works. Even though the results are not directly comparable, because different annotation styles were used, we suppose that the annotation unification across the treebanks in Universal Dependencies should improve the transfer parser scores. [14] presented 61.7 % of averaged accuracy over 8 languages. On the same languages, our transfer parser on UD reached 70.1 %. When compared to [20], we experimented with 23 common languages, our average score on them is 62.5 %, Rosa's is 56.6 %. The higher attachment scores in our experiments confirms that the annotations in UD treebanks are more unified and serve better for transferring between languages.

9   Conclusions

We used the Universal Dependencies treebank collection to test delexicalized transfer parsers and unsupervised dependency parser enriched by external *attach* and *stop* prior probabilities. We found that whereas the MST delexicalized transfer parser is better in average, our minimally supervised parser performs better on many non-Indo-European languages and therefore can be suitable to parse often low-resourced exotic languages, for which treebanks do not exist.

Acknowledgments. This work has been supported by the grant 14-06548P of the Czech Science Foundation.

---

[10] Danish is the only exception.

References

1. Blunsom, P., Cohn, T.: Unsupervised induction of tree substitution grammars for dependency parsing. In: Proceedings of the 2010 Conference on Empirical Methods in Natural Language Processing, pp. 1204–1213. EMNLP 2010. Association for Computational Linguistics, Stroudsburg (2010)
2. Buchholz, S., Marsi, E.: CoNLL-X shared task on multilingual dependency parsing. In: Proceedings of the Tenth Conference on Computational Natural Language Learning, pp. 149–164. CoNLL-X 2006. Association for Computational Linguistics, Stroudsburg (2006)
3. Cerisara, C., Lorenzo, A., Kral, P.: Weakly supervised parsing with rules. In: Interspeech 2013, Lyon, France, pp. 2192–2196 (2013)
4. Gilks, W.R., Richardson, S., Spiegelhalter, D.J.: Markov chain Monte Carlo in practice. Interdisciplinary Statistics. Chapman & Hall, London (1996)
5. Hajič, J., Hajičová, E., Panevová, J., Sgall, P., Pajas, P., Štěpánek, J., Havelka, J., Mikulová, M.: Prague Dependency Treebank 2.0. CD-ROM, Linguistic Data Consortium, LDC Catalog No.: LDC2006T01, Philadelphia (2006)
6. Headden III, W.P., Johnson, M., McClosky, D.: Improving unsupervised dependency parsing with richer contexts and smoothing. In: Proceedings of Human Language Technologies: The 2009 Annual Conference of the North American Chapter of the Association for Computational Linguistics. NAACL 2009, pp. 101–109. Association for Computational Linguistics, Stroudsburg (2009)
7. Klein, D., Manning, C.D.: Corpus-based induction of syntactic structure: models of dependency and constituency. In: Proceedings of the 42nd Annual Meeting on Association for Computational Linguistics. ACL 2004. Association for Computational Linguistics, Stroudsburg (2004)
8. Marecek, D.: Multilingual unsupervised dependency parsing with unsupervised POS tags. In: Sidorov, G., et al. (eds.) MICAI 2015. LNCS, vol. 9413, pp. 72–82. Springer, Heidelberg (2015). doi:10.1007/978-3-319-27060-9_6
9. Mareček, D., Straka, M.: Stop-probability estimates computed on a large corpus improve unsupervised dependency parsing. In: Proceedings of the 51st Annual Meeting of the Association for Computational Linguistics (Volume 1: Long Papers), pp. 281–290. Association for Computational Linguistics, Sofia, August 2013
10. de Marneffe, M.C., Dozat, T., Silveira, N., Haverinen, K., Ginter, F., Nivre, J., Manning, C.D.: Universal stanford dependencies: a cross-linguistic typology. In: Proceedings of the 9th Conference on Language Resources and Evaluation (LREC) (2014)
11. de Marneffe, M.C., Manning, C.D.: The stanford typed dependencies representation. In: Coling 2008: Proceedings of the Workshop on Cross-Framework and Cross-Domain Parser Evaluation. CrossParser 2008, pp. 1–8. Association for Computational Linguistics, Stroudsburg (2008)
12. Mcdonald, R., Nivre, J., Quirmbach-brundage, Y., Goldberg, Y., Das, D., Ganchev, K., Hall, K., Petrov, S., Zhang, H., Tckstrm, O., Bedini, C., Bertomeu, N., Lee, C.J.: Universal dependency annotation for multilingual parsing. In: Proceedings of ACL 2013 (2013)
13. McDonald, R., Pereira, F., Ribarov, K., Hajič, J.: Non-projective dependency parsing using spanning tree algorithms. In: Proceedings of Human Langauge Technology Conference and Conference on Empirical Methods in Natural Language Processing (HTL/EMNLP), Vancouver, BC, Canada, pp. 523–530 (2005)

14. McDonald, R., Petrov, S., Hall, K.: Multi-source transfer of delexicalized dependency parsers. In: Proceedings of the Conference on Empirical Methods in Natural Language Processing. EMNLP 2011, pp. 62–72. Association for Computational Linguistics, Stroudsburg (2011)

15. Naseem, T., Chen, H., Barzilay, R., Johnson, M.: Using universal linguistic knowledge to guide grammar induction. In: Proceedings of the 2010 Conference on Empirical Methods in Natural Language Processing. EMNLP 2010, pp. 1234–1244. Association for Computational Linguistics, Stroudsburg (2010)

16. Nivre, J.: Non-projective dependency parsing in expected linear time. In: Su, K.Y., Su, J., Wiebe, J. (eds.) ACL/IJCNLP, pp. 351–359. The Association for Computer Linguistics, Stroudsburg (2009)

17. Nivre, J., Hall, J., Kübler, S., McDonald, R., Nilsson, J., Riedel, S., Yuret, D.: The CoNLL 2007 shared task on dependency parsing. In: Proceedings of the CoNLL Shared Task Session of EMNLP-CoNLL 2007, pp. 915–932. Association for Computational Linguistics, Prague, June 2007

18. Nivre, J., de Marneffe, M.C., Ginter, F., Goldberg, Y., Hajič, J., Manning, C., McDonald, R., Petrov, S., Pyysalo, S., Silveira, N., Tsarfaty, R., Zeman, D.: Universal dependencies v1: a multilingual treebank collection. In: Proceedings of the 10th International Conference on Language Resources and Evaluation (LREC 2016). European Language Resources Association, Portorož (2016)

19. Petrov, S., Das, D., McDonald, R.: A universal part-of-speech tagset. In: Proceedings of the Eight International Conference on Language Resources and Evaluation (LREC 2012). European Language Resources Association (ELRA), Istanbul, May 2012

20. Rosa, R.: Multi-source cross-lingual delexicalized parser transfer: Prague or Stanford? In: Proceedings of the Third International Conference on Dependency Linguistics (Depling 2015), pp. 281–290. Uppsala University, Uppsala (2015)

21. Rosa, R., Mašek, J., Mareček, D., Popel, M., Zeman, D., Žabokrtský, Z.: HamleDT 2.0: thirty dependency treebanks stanfordized. In: Proceedings of the Ninth International Conference on Language Resources and Evaluation (LREC-2014), Reykjavik, Iceland, May 26–31, 2014, pp. 2334–2341 (2014)

22. Spitkovsky, V.I., Alshawi, H., Chang, A.X., Jurafsky, D.: Unsupervised dependency parsing without gold part-of-speech tags. In: Proceedings of the 2011 Conference on Empirical Methods in Natural Language Processing (EMNLP 2011) (2011). pubs/goldtags.pdf

23. Spitkovsky, V.I., Alshawi, H., Jurafsky, D.: Punctuation: making a point in unsupervised dependency parsing. In: Proceedings of the Fifteenth Conference on Computational Natural Language Learning (CoNLL-2011) (2011)

24. Spitkovsky, V.I., Alshawi, H., Jurafsky, D.: Three dependency-and-boundary models for grammar induction. In: Proceedings of the 2012 Conference on Empirical Methods in Natural Language Processing and Computational Natural Language Learning (EMNLP-CoNLL 2012) (2012). pubs/dbm.pdf

25. Zeman, D.: Reusable tagset conversion using tagset drivers. In: Proceedings of the Sixth International Conference on Language Resources and Evaluation (LREC 2008). European Language Resources Association (ELRA), Marrakech (May 2008). http://www.lrec-conf.org/proceedings/lrec2008/

26. Zeman, D., Dušek, O., Mareček, D., Popel, M., Ramasamy, L., Štěpánek, J., Žabokrtský, Z., Hajič, J.: HamleDT: harmonized multi-language dependency treebank. Lang. Resour. Eval. **48**(4), 601–637 (2014)

27. Zeman, D., Mareček, D., Popel, M., Ramasamy, L., Štěpánek, J., Žabokrtský, Z., Hajič, J.: HamleDT: to Parse or not to parse? In: Proceedings of the Eight International Conference on Language Resources and Evaluation (LREC 2012). European Language Resources Association (ELRA), Istanbul (2012)
28. Zeman, D., Resnik, P.: Cross-language parser adaptation between related languages. In: IJCNLP 2008 Workshop on NLP for Less Privileged Languages, pp. 35–42. Asian Federation of Natural Language Processing. International Institute of Information Technology, Hyderabad (2008)

Unsupervised Morphological Segmentation Using Neural Word Embeddings

Ahmet Üstün[1(✉)] and Burcu Can[2(✉)]

[1] Cognitive Science Department, Informatics Institute,
Middle East Technical University (ODTÜ), Ankara 06800, Turkey
ustun.ahmet@metu.edu.tr
[2] Department of Computer Engineering, Hacettepe University,
Beytepe, Ankara 06800, Turkey
burcucan@cs.hacettepe.edu.tr

Abstract. We present a fully unsupervised method for morphological segmentation. Unlike many morphological segmentation systems, our method is based on semantic features rather than orthographic features. In order to capture word meanings, word embeddings are obtained from a two-level neural network [11]. We compute the semantic similarity between words using the neural word embeddings, which forms our baseline segmentation model. We model morphotactics with a bigram language model based on maximum likelihood estimates by using the initial segmentations from the baseline. Results show that using semantic features helps to improve morphological segmentation especially in agglutinating languages like Turkish. Our method shows competitive performance compared to other unsupervised morphological segmentation systems.

Keywords: Morphology · Semantics · Neural representation of speech and language · Morphological segmentation · Unsupervised learning · Word embeddings

1 Introduction

Morphological analysis is the heart of nearly all natural language processing tasks, such as sentiment analysis, machine translation, information retrieval etc. Such natural language processing tasks become infeasible without any morphological analysis. One reason is the sparsity that is the result of a high number of word forms which introduces out-of-vocabulary (OOV). Morphological segmentation is a way to deal with language sparsity by introducing the common segments within the words rather than dealing with word forms (having multiple morphemes). Hankamer [7] claims that the number of word forms is infinite in agglutinating languages like Turkish.

Morphology is strongly connected to other linguistic levels, such as syntax and semantics. Although this connection has been addressed many times in the literature, semantic features have been scarcely used in morphological segmentation.

© Springer International Publishing AG 2016
P. Král and C. Martín-Vide (Eds.): SLSP 2016, LNAI 9918, pp. 43–53, 2016.
DOI: 10.1007/978-3-319-45925-7_4

In this paper, we propose an unsupervised method to morphological segmentation that integrates morphotactics with semantics. Our method makes use of semantic similarity between words in order to learn the segmentation points. For example, *book-booking*, *booking-bookings*, *book-booker* are all semantically similar, which gives a clue while segmenting the word *bookings* into *book*, *ing* and *s*; *booker* into *book* and *er*. Using the semantic features, we infer the segmentation points and we use the potential segmentation points in order to model the morphotactic rules for the morpheme transitions. This forms the baseline model in this paper.

We obtain the semantic features of words from the word embeddings, which are learned by a two-layer neural networks [11]. Thus, word meanings are represented in a low-dimensional vector space. We use the cosine similarity between the word embeddings in order to measure the semantic similarity.

Morphotactics of the language is modeled by maximum likelihood estimate based on the initial segmentations of words obtained from the baseline model. Hence, we integrate semantics and morphotactics within the same model.

The paper is organized as follows: Section 2 addresses the related work on unsupervised morphological segmentation, Sect. 3 describes the mathematical model where Sect. 3.2 describes our baseline model based on semantic similarity and Sect. 3.3 describes the bigram model based on maximum likelihood estimates, and Sect. 4 presents the experiment results compared to other systems.

2   Related Work

Many of the unsupervised morphological segmentation systems are based on word-level orthographic patterns. Goldwater et al. [6] present a two-stage Bayesian model, where morpheme types are drawn from a multinomial distribution (i.e. *generator*) and tokens are generated by a Pitman-Yor process to have a power-law distribution over word frequencies (i.e. *adaptor*).

Morfessor family involves different unsupervised morphological segmentation models. Morfessor Baseline [4] is based on Minimum Description Length (MDL) model and aims to find the lexicon of morphemes that will minimize the corpus length.

Morfessor CatMAP (Creutz and Lagus [5]) performs morphological segmentation using a lexicon of morpheme types and corpus that involves the word tokens to be segmented. Morpheme features such as frequency or perplexity are used as ortographic features. Morphotactics is also modeled with a maximum a posteriori framework (MAP) using Hidden Markov Models (HMMs) for the morpheme transitions. Words are represented as HMMs with four hidden variables: stem, prefix, suffix, and non-morph as for the noisy morphemes. The results show that modeling morphotactics improves morphological segmentation. Our method is similar to Morfessor in term of handling the morphotactics. This is one of the intuitions that we follow in this paper.

Connection between morphology and syntax has been previously addressed in the literature. Can and Manandhar [1] use syntactic categories obtained by context distribution clustering of Clark [2] for finding morphological paradigms. Lee et al. [9] incorporates part-of-speech information with the morphological segmentation on Arabic. Results show that using syntactic context also improves morphological segmentation.

Connection between morphology and semantics has also been addressed in the literature. Schone and Jurafsky [14] use Latent Semantic Analysis (LSA) to measure the semantic similarity between similar surface forms of the words. This comes from the idea that words which are morphologically derived from each other also semantically similar. For example, *car* and *care* are semantically not similar, therefore they cannot be morphologically derived from each other.

We are also inspired by Schone and Jurafsky [14] in this paper. We learn the segmentation points of the words using the semantic similarity between word forms. Differently from Schone and Jurafsky [14], we use neural word embeddings with a low dimensional vector space to measure semantic similarity.

Narasimhan et al. [12] present a log-linear model, where semantic information is used as feature in the model. Semantic features are obtained from vector space representations and cosine similarity computed between word pairs, which is similar to our model presented in this paper.

Soricut and Och [15] capture morphological rules and morphological paradigms from semantic features obtained from a vector space of words.

Our model resembles the works by Schone and Jurafsky [14] and Narasimhan et al. [12] since we also use the semantic similarity to learn morphological units in words. Progress in learning word embeddings from neural networks have made the semantic information available in computational linguistics in the recent years. Here, we also adopt the semantic information obtained from neural word embeddings in this paper.

3   Morphological Segmentation with Word Embeddings

3.1   Model Overview

Our model consists of three steps that are performed sequentially:

1. We obtain word embeddings from a raw corpus in order to collect semantic information. Skip-gram [11] model is used to build word vectors in 200 dimensional vector space.
2. Words are initially segmented recursively by measuring the semantic similarity between substrings in each word, which forms the baseline model. Semantic similarity is computed by using the word embeddings obtained in the first step.
3. We model the morphotactics by adopting a bigram language model with a maximum likelihood estimate.

Details of three steps are explained below.

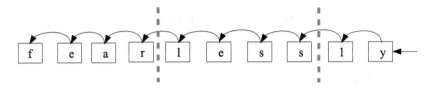

Fig. 1. The illustration that shows the segmentation of the word *fearlessly*. The segmentation is performed from the right end of the word by stripping off letters depending on the manually set semantic similarity.

Table 1. The cosine similarity between the substrings of the word *fearlessly*.

	Word	Remaining substring	Cosine similarity	Segmentation
1	fearlessly	fearlessl	−1	fearlessly
2	fearlessly	fearless	0.34	fearless-ly
3	fearless	fearles	0.14	fearless-ly
4	fearless	fearle	−1	fearless-ly
5	fearless	fear	0.26	fear-less-ly
6	fear	fea	−1	fear-less-ly
7	fear	fe	−1	**fear-less-ly**

3.2   Morphological Segmentation Using Semantic Similarity

Morphological affixation, either inflection or derivation, is strongly connected to semantic derivation that leads to different word forms which are all semantically related with each other. This semantic relation constitutes one side of creativity in language as a cognitive phenomenon. For example, English word *fear* has a close semantic relation with *fearless* and *fearlessly*. The same also holds for inflection. The words *car* and *cars*, *study* and *studied* have got a very close semantic relation.

The initial segmentation is based on this semantic relation between different word forms which are all derived from the same root. We detect the segmentation boundaries based on the semantic similarity between substrings of a word. Semantic similarity is calculated by cosine similarity between word embeddings of the substrings in n dimensional space as follows:

$$\cos(v(w_1), v(w_2)) = \frac{v(w_1) \cdot v_i(w_2)}{\|v(w_1)\| \cdot \|v_i(w_2)\|} \tag{1}$$

$$= \frac{\sum_{i=1}^{n} v_i(w_1) \cdot v_i(w_2)}{\sqrt{\sum_{i=1}^{n} v_i(w_1)^2} \cdot \sqrt{\sum_{i=1}^{n} v_i(w_2)^2}} \tag{2}$$

where $v(w_1)$ denotes the word embedding for of the word w_1, $v(w_2)$ denotes the word embedding for the word w_2 and $v_i(w_1)$ is the ith item in $v_i(w_1)$.

In each iteration of the algorithm, one letter is stripped from the right end of the word. If the cosine similarity between the remaining substring of the

Algorithm 1. Baseline segmentation algorithm that detects potential morpheme boundaries based on cosine similarity.

```
 1: procedure SEMANTIC_PARSING(word)
 2:     boundaryList ← ∅
 3:     threshold ← d
 4:     wordLen ← LEN(word)
 5:     charNo ← 1
 6:     oldWord ← word
 7:     counter ← wordLen
 8:     while counter > 1 do
 9:         suffix ← LAST_CHAR(charNo, oldWord).
10:         newWord ← FIRST_CHAR(wordLen − charNo, oldWord).
11:         distance ← COSINE(newWord, oldWord)
12:         if distance > threshold then
13:             boundaryList ← PUT(suffix)
14:             oldWord ← newWord
15:             charNo ← 1
16:             wordLen ← LEN(oldWord)
17:         else
18:             charNo ← charNo + 1
19:         counter ← counter − 1
20:     return boundaryList
```

word and the substring before stripping a letter is above a threshold value, the segmentation point is accepted. Otherwise, another letter is stripped from the end of the word and concatenated with the letter which has been stripped in the previous iteration. Similarly, a segmentation point is introduced or not depending on the cosine similarity between the remaining substring and the substring with the stripped letters. The entire word is checked from right to left by detecting potential segmentation points until all the letters of the word are stripped. An example segmentation is given in Fig. 1 and corresponding cosine similarity values for the same word are given in Table 1.

We use a manually set threshold for the cosine similarity d to decide whether a semantic relation is protected between word forms or not. The entire procedure is given in Algorithm 1. $LEN(word)$ is the number of letters in the word, $LAST\_CHAR(charNo, oldWord)$ returns the rightmost substring that has a number of $charNo$ letters in $oldWord$, $FIRST\_CHAR(charNo, oldWord)$ returns the leftmost substring that has a number of $charNo$ letters in $oldWord$, and $COSINE(newWord, oldWord)$ calculates cosine similarity between $newWord$ and $oldWord$ with respect to Eq. 1.

Words that are not found in the corpus do not have a word embedding. If an unseen word is encountered, semantic similarity is assigned -1 and no segmentation is suggested for these words. This is the case in the baseline model and a segmentation is suggested for the unseen words in the maximum likelihood model, which is described in the next section.

3.3   Modeling Morphotactics with ML Estimate

Morphotactics involves a set of rules that define how morphemes can be attached to each other [16]. In agglutinating languages like Turkish, Finnish and Hungarian, concatenation of morphemes plays an important role in morphological generation that builds different words and word forms.

We use maximum likelihood estimation to build a bigram language model for morpheme transitions. Unigram probabilities are used for the first segments, namely the roots, and bigram transition probabilities are used for the suffix ordering. Maximum likelihood model is given as follows:

$$\arg\max_{w=m_0+\cdots+m_N \in W} P(w = m_0 + m_1 + \cdots + m_N) = p(m_0) \prod_{i=1}^{N} p(m_i|m_{i-1}) \quad (3)$$

where w is a word in corpus W and m_i refers to the ith morpheme in w that consists of N morphemes. For both unigram and bigram probabilities we use the maximum likelihood estimates that are obtained from the initial segmentations in the baseline model:

$$p(m_0) = \frac{n(m_0)}{K} \quad (4)$$

where $n(m_0)$ is the frequency of m_0 and K is the total frequency of morphemes in the segmented corpus. The bigram probability is calculated as follows:

$$p(m_i|m_{i-1}) = \frac{n(<m_i, m_{i-1}>)}{M} \quad (5)$$

where $n(<m_i, m_{i-1}>)$ is the frequency of the bigram $<m_i, m_{i-1}>$ and M is the total frequency of the bigram types in the segmented corpus.

An end-of-word symbol is added at the end of each word as a morpheme in order to assign a probability for the last morpheme being the final morpheme of a word. For example, the probability of the last morpheme in *fearlessly$* being ly is computed by $p(\$|ly)$.

While calculating the bigram probabilities, root morpheme is changed to a start symbol in order to remove any dependencies between the root and the first morpheme. For example, the probability of the first morpheme after the root being *less* is computed as follows:

$$p(less|S) = \frac{n(<S, less>)}{M} \quad (6)$$

where $n(<S, less>$ is the frequency of *less* seen as the first morpheme just after the root in the corpus.

In order to find the segmentation points in a word, all possible segmentations are obtained and the segmentation with the maximum likelihood probability regarding Eq. 3 is selected as the final segmentation of the word. Viterbi algorithm is used for finding the segmentation having the maximum likelihood. We apply Laplace smoothing with additive number 1 to overcome the sparsity problem.

Table 2. Corpora size for English and Turkish

	English	Turkish
Word embeddings	129M	361M
Semantic parsing and ML estimation [8]	878K	617K
Development [8]	694	763
Test and evaluation [8]	1050	1760

Table 3. F1-measure on Turkish development set for different cosine similarity threshold values

Threshold (d)	Semantic parsing (%)	Full model (%)
0.15	40.51	47.51
0.25	37.42	47.82
0.35	30.16	43.58
0.45	25.14	39.95

4 Experiments and Results

4.1 Data

We used publicly available Morpho Challenge [8] data for training and testing. We tested our model on two languages: English and Turkish. The English dataset consists of 878,034 words and the Turkish dataset consists of 617,298 words with their frequencies in the corpus. The frequency is used for the maximum likelihood estimates of the morphemes in the model.

In order to learn the word embeddings for the words, we used manually collected data both for English and Turkish. The English corpus consists of 129 million word tokens and 218 thousand word types, Turkish corpus consists of 361 million word tokens and 725 thousand word types.

The evaluation was performed on Morpho Challenge [8] gold standard data. Corpora details are given in Table 2. For the evaluation, a number of word pairs are sampled from the results that share at least one common morpheme. For each word pair that indeed share a common morpheme according to the gold standard segmentations, one point is given. Total number of points is divided by the total number of word pairs. This gives the precision. The same is applied for recall by sampling word pairs that share at least one common morpheme from the gold standard segmentations and checked from the results whether they indeed share a common morpheme.

During semantic parsing step in the baseline method, we assign cosine similarity threshold $d = 0.25$ to decide whether two substrings of a word are semantically similar or not. In order to manually assign the threshold value d, we perform both semantic parsing and the full model on the Turkish development corpora [8]. Results are given in Table 3 for different values of the cosine similarity threshold value.

Table 4. Morpheme segmentation results on Turkish corpora

Model	Precision (%)	Recall (%)	F1-measure (%)
Morfessor CatMap [3]	79.38	31.88	45.49
Full model	50.70	40.07	44.76
Morpho Chain [12]	69.63	31.73	43.60
Aggressive comp. [10]	55.51	34.36	42.45
Semantic parsing	61.82	25.42	36.03
Iterative comp. [10]	68.69	21.44	32.68
Morfessor baseline [4]	87.35	18.03	29.89
Nicolas [13]	79.02	19.78	31.64
Base inference [10]	72.81	16.11	26.38

Table 5. Morpheme segmentation results on English corpora

Model	Precision (%)	Recall (%)	F1-measure (%)
Morfessor baseline [4]	66.30	41.28	50.88
Semantic parsing	64.85	37.75	47.72
Full model	62.79	35.40	45.28
Morfessor CatMap [3]	64.44	34.34	44.81

Mikolov's [11] word2vec tool and its open source Java implementation Deeplearning4J [17] are used to obtain vword embeddings for the words.

4.2 Evaluation and Results

We compare our model with Morfessor Baseline [4] and Morfessor Categories MAP [3] since these are well known unsupervised morphological segmentation systems and they perform well in both English and Turkish. However, neither of the models use any semantic information for the segmentation task.

We run Morfessor Baseline, Morfessor CatMAP and our model on the same training and test corpora [8]. Turkish results are given in Table 4 with the other Morpho Challenge 2010 [8] participants. English results are given in Table 5 compared to Morfessor Baseline and Morfessor CatMAP. We also provide our baseline model results that are obtained by adopting only semantic parsing given in Sect. 3.2.

Our model comes the second out of nine systems on Turkish with a F-measure 44.76 %. Our baseline results are also promising and outperforms Morfessor Baseline. It is promising to see that using semantic similarity between morphologically related words through only the word embeddings is sufficient for a simple morphological segmentation.

Table 6. Correct and incorrect examples of English results

Correct segmentations	Incorrect segmentations
vouch-safe-d	cen-tr-alize-d
dictator-ial	ni-hil-ist-ic
help-less-ness	su-f-fix-es
rational-ist	ba-ti-ste
express-way	sh-o-gun
flow-chart	el-e-v-ation-s
drum-head-s	im-pe-rsonator-s

Table 7. Correct and incorrect examples of Turkish results

Correct segmentations	Incorrect segmentations
patlıcan-lar-ı	tiy-at-ro-lar-da
su-lar-da-ki	gaze-t-e-ci-ydi
balkon-lar-da	sipari-ş-ler-i-n-iz
parti-si-ne	gelişti-ril-ir-ken
varis-ler-den	anla-ya-mıyo-r-du-m
entari-li-nin	uygu-lama-sı-nda-n
üye-ler-i-dir-ler	veri-tabanları-yla

Our model also gives competitive results on English when compared to Morfessor Baseline [4] and Morfessor Categories MAP [3] with a F-measure 45.28 %.

One of the common errors arise from highly inflected words, which are prone to over-segmentation in the model. Some morphemes are substrings of other morphemes and therefore there is a semantic relation between these substrings. Therefore, two different morphemes are introduced instead of a single morpheme in these cases. For example, *lar* is a valid morpheme in Turkish; *la* and *r* are also valid morphemes in Turkish. Therefore, the word *kitaplar* is segmented into *kitap*, *la* and *r* since *kitapla* and *kitaplar* are both valid words and semantically related to each other.

Moreover, it is hard to capture derivational suffixes due to the semantic changes that are introduced by derivation. However, overall success of the baseline method is very promising compared to the methods that work only in the ortographic word level.

Examples of both correct segmentations and incorrect segmentations in English and in Turkish that are obtained by our full model are given in Tables 6 and 7.

5   Conclusion and Future Work

We present a novel probabilistic model for unsupervised morphological segmentation that adopts the word embeddings in order to measure semantic similarity between word forms which are built from the same root. The results show that incorporating semantic similarity helps finding morphologically related words that leads to find the morpheme boundaries in a simple approach.

We adopt a bigram language model by using maximum likelihood estimate in order to learn morphotactics in the bigram level. We plan to use a better language model for the morpheme transitions in the future.

Acknowledgements. This research was supported by TUBITAK (The Scientific and Technological Research Council of Turkey) grant number 115E464.

References

1. Can, B., Manandhar, S.: Clustering morphological paradigms using syntactic categories. In: Peters, C., Di Nunzio, G.M., Kurimo, M., Mandl, T., Mostefa, D., Peñas, A., Roda, G. (eds.) CLEF 2009. LNCS, vol. 6241, pp. 641–648. Springer, Heidelberg (2010)
2. Clark, A.: Inducing syntactic categories by context distribution clustering. In: Proceedings of 2nd Workshop on Learning Language in Logic and 4th Conference on Computational Natural Language Learning, ConLL 2000, vol. 7, pp. 91–94. Association for Computational Linguistics, Stroudsburg (2000)
3. Creutz, M., Lagus, K.: Inducing the morphological lexicon of a natural language from unannotated text. In: Proceedings of International and Interdisciplinary Conference on Adaptive Knowledge Representation and Reasoning (AKRR 2005), pp. 106–113 (2005)
4. Creutz, M., Lagus, K.: Unsupervised morpheme segmentation and morphology induction from text corpora using morfessor 1.0. Technical report A81 (2005)
5. Creutz, M., Lagus, K.: Unsupervised models for morpheme segmentation and morphology learning. ACM Trans. Speech Lang. Process. **4**, 3:1–3:34 (2007)
6. Goldwater, S., Griffiths, T.L., Johnson, M.: Interpolating between types and tokens by estimating power-law generators. In: Advances in Neural Information Processing Systems, vol. 18, p. 459 (2006)
7. Hankamer, J.: Finite state morphology and left to right phonology. In: Proceedings of 5th West Coast Conference on Formal Linguistics, January 1986
8. Kurimo, M., Lagus, K., Virpioja, S., Turunen, V.T.: Morpho Challenge 2010, June 2011. http://research.ics.tkk.fi/events/morphochallenge2010/. Accessed 4 Jul 2016
9. Lee, Y.K., Haghighi, A., Barzilay, R.: Modeling syntactic context improves morphological segmentation. In: Proceedings of 15th Conference on Computational Natural Language Learning, CoNLL 2011, pp. 1–9. Association for Computational Linguistics, Stroudsburg (2011)
10. Lignos, C.: Learning from unseen data. In: Kurimo, M., Virpioja, S., Turunen, V., Lagus, K. (eds.) Proceedings of Morpho Challenge 2010 Workshop, pp. 35–38. Aalto University, Espoo (2010)
11. Mikolov, T., Chen, K., Corrado, G., Dean, J.: Efficient estimation of word representations in vector space (2013). CoRR arXiv:abs/1301.3781

12. Narasimhan, K., Barzilay, R., Jaakkola, T.S.: An unsupervised method for uncovering morphological chains. Trans. Assoc. Comput. Linguist. (TACL) **3**, 157–167 (2015)
13. Nicolas, L., Farré, J., Molinero, M.A.: Unsupervised learning of concatenative morphology based on frequency-related form occurrence. In: Kurimo, M., Virpioja, S., Turunen, V., Lagus, K. (eds.) Proceedings of Morpho Challenge 2010 Workshop, pp. 39–43. Aalto University, Espoo (2010)
14. Schone, P., Jurafsky, D.: Knowledge-free induction of inflectional morphologies. In: Proceedings of 2nd Meeting of the North American Chapter of the Association for Computational Linguistics on Language Technologies, NAACL 2001, pp. 1–9. Association for Computational Linguistics, Stroudsburg (2001)
15. Soricut, R., Och, F.: Unsupervised morphology induction using word embeddings. In: Human Language Technologies: The 2015 Annual Conference of the North American Chapter of the ACL, pp. 1627–1637 (2015)
16. Sproat, R.W.: Morphology and Computation. MIT Press, Cambridge (1992)
17. Team, D.D.: Deeplearning4j: Open-Source Distributed Deep Learning for the JVM, Apache Software Foundation License 2.0, May 2016. http://deeplearning4j.org/

Speech

Statistical Analysis of the Prosodic Parameters of a Spontaneous Arabic Speech Corpus for Speech Synthesis

Ikbel Hadj Ali and Zied Mnasri[✉]

Signal, Image and Technology of Information Laboratory,
Ecole Nationale d'Ingenieurs de Tunis-ENIT,
University Tunis El-Manar, Tunis, Tunisia
ikbelhadjali@gmail.com, zied.mnasri@enit.rnu.tn

Abstract. In this paper, we present the analysis of a normalized and representative spontaneous Arabic speech corpus with labeling and annotation, in order to provide a complete library of voice segments and their respective prosodic parameters at different levels (phonemic or syllabic). A statistical analysis was conducted afterwards to determine and normalize the distribution of the collected data. The obtained results were then compared to those of a prepared Arabic speech corpus, in order to determine the characteristics of each kind of speech corpus and its suitable application area.

Keywords: Phonetic analysis · Spontaneous speech · Arabic speech corpus · Prosodic parameters · Statistical distribution

1 Introduction

Speech processing applications are catching up the continuous development of multimedia. Speech enhancement, recognition, synthesis can be found in all the recent technologies of mobile devices and computers. On the other hand, the Arabic speaking world is a large multimedia market, with at least 300 million potential users. Therefore, there is an urgent need to adapt the new speech processing techniques to Arabic.

Arabic is a Semitic language spoken today by more than 300 million people worldwide and is the official language of 22 countries, located in North Africa, the African horn and the middle-east, in addition to the Arabic Diaspora and the Arabic speaking Muslims. Due to this large geographic extension, Arabic dialects are very different, and sometimes it is hard for two Arabic speaking individuals, from different parts of the Arabic world to understand each other, unless they both speak the standard version of Arabic. This standard version is the official language for literature, journalism and science in these countries. However, dialects difference influence the way of articulation and alters the prosodic parameters very significantly.

P. Král and C. Martín-Vide (Eds.): SLSP 2016, LNAI 9918, pp. 57–67, 2016.
DOI: 10.1007/978-3-319-45925-7_5

The prosodic parameters, i.e. F0, duration and intensity are the phonetic, or the physical manifestation of the phonological features of speech, i.e. intonation, accentuation and energy [7]. Actually, the way of articulation and the context of speech have a great influence on these parameters. For instance, a speaker spelling radio news realizes different values of F0, duration and intensity than somebody saying the same text spontaneously. Therefore, automatic speech processing systems have to take care of the nature of speech corpus, whether spontaneous or prepared, before modeling the prosodic parameters. This paper is organized as follows: the first section introduces the prosodic parameters, the second section presents the characteristics of the spontaneous Arabic speech corpus used in this study, the third one describes the operations of analysis, annotation and labeling of this corpus, whereas the fourth section shows the statistical analysis of the extracted parameters in a comparison with a previously studied prepared corpus.

2   Statistical Modeling of Prosodic Parameters

Prosodic parameters modeling is necessary to predict the values of duration, F0 and intensity from the text and or the context. To fulfill that, a variety of models of different types have been set, whether phonological, analytic or statistical models. Recently, statistical learning tools, like neural networks, SVM and HMM have been widely used to model prosodic parameters [5,9,13]. However, to achieve good statistical modeling, the training data should be normally distributed [11]. Otherwise, a normalization process should be undertaken before starting the training. Hence, both the actual distribution of each prosodic parameter, and its normalization transform should be studied, for each type of speech corpus, i.e. spontaneous or prepared, in order to decide which corpus to use in training the statistical models of prosodic parameters.

Statistical models require a standardized database containing representative prosodic parameters of the different levels of units, i.e. phonemes, syllables, words, as well as the characteristic descriptors that will be trained in order to generate a model allowing to predict the prosodic parameters from the textual and the contextual features. First, the prosodic parameters have to be extracted, at different levels. For each segment, the textual/contextual features are extracted as well. In a second step, the features are classified into categorical and ordinal values, and then normalized to a certain range. On the opposite side, the targets, i.e. the prosodic parameters are statistically studied to determine the distribution of each. If the distribution is not Gaussian, then a transformation is looked for to normalize it.

This study is based on both levels of speech elementary segments, i.e. the phoneme and the syllable. Actually, both levels were adopted by phoneticians to model prosody. The syllable is said to be the smallest generic frame duration, where the phonemes included inside are stretched by the same factor [5], whereas the phoneme is said to be the elementary unit of speech [12], and the carrier of the micro-prosodic variations, which is mainly related to pitch. Then, the use of larger units to model the prosody may lead to over-smoothed parameters, which may cause an envelope effect in speech synthesis for example [1] (Fig. 1).

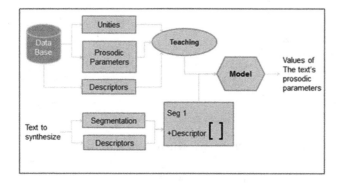

Fig. 1. Prosodic parameters prediction system

3 Spontaneous Arabic speech database

3.1 Specificities of Spontaneous Arabic Speech

The Arabic language is considered as a difficult language to master in the field of natural language processing due to its complicated morphological and syntactic properties. Research for automatic processing of the Arabic word began around the 1970s with early work concerning the lexicon and morphology [2]. Amongst the particularities of the Arabic language we cite:

- Arabic is written and read from right to left.
- There are no alphabet characters for short vowels, but diacritic signs, which are rarely mentioned in Arabic printed documents.
- The shape of an alphabetic character depends on its position in the word and its diacritic sign.

The alphabet of the Arabic language has 28 letters which are consonants, although three of them are also used as characters for long vowels ([a:], [u:] and [i:]). We consider for our research, based on SAMPA code, that the Arabic alphabet has 28 consonants and 6 vowels (3 short vowels [a], [u], [i], and 3 long ones [a:], [u:], [i:]). With SAMPA code, consonants of the Arabic language are classified into six subclasses (cf. Table 1).

The Arabic language has six syllable types classified according to their length or openness. A syllable is called open (respectively closed) if it ends by a vowel (V) (respectively by a consonant (C)). (cf. Table 2).

The spontaneous speech is characterized by some special phonological phenomena such as: pauses, false starts, repetitions, hesitations, ungrammaticality, intonation and rhythm as well as the emotional state of the speaker. Meanwhile, these features are absent in the prepared speech corpus which ensures clear intonation with phonetically balanced sentences [4].

Table 1. Arabic phonemes in SAMPA code

	Voiced	Unvoiced
Plosive	[b]/ [d.]/ [d]	[t.]/ [t]/ [k]/ [q]/ [?]
Fricatives	[z.]/ [D]/ [z]/ [G]/ [H]	[f]/ [T]/ [s.]/ [s]/ [S]/ [x]/ [X]/ [h]
Affricate	[Z]	
Nasal	[n]/ [m]	
Lateral	[l]	
Vibrant	[r]	
Semi-vowel	[j]/ [w]	

Table 2. Arabic syllables classification

	Open	Closed
Short	CV/ CVV	CVC
Long		CVVC/ CVCC/ CVVCC

3.2   Reference Corpus

The selection of records to be used in the speech database is the most important step. The database must be large enough to contain all possible phonetic sequences that can appear in any sort of linguistic contexts. The recording conditions of spontaneous speech should also keep the natural aspect of conversation. Therefore, we chose to record natural extracts spoken by a male speaker and belonging to several kinds or conditions of production (radio news, radio interview...etc.). Our corpus includes speech samples for a total of 300 s in wav format. It includes affirmative and interrogative utterances, with an amount of 1127 syllables and 2718 phonemes. The transcription of each audio file is converted from graphemes to phonemes based on SAMPA code. To ensure a good comparison between the spontaneous and the prepared speech, the nature and the context of the spontaneous database was similar to the prepared corpus. Actually the latter, processed in a previous work [10], was recorded in a phonetically insulated studio by a male speaker. It includes 191 affirmative and interrogative sentences containing 1556 syllables and 3461 phonemes.

3.3   Segmentation, Labeling and Annotation

The next step is the Segmentation, annotation and labeling of the spontaneous corpus of the Arabic speech using PRAAT tool [3]. Actually, segmentation was done manually using both the waveform and the spectrogram to set the phoneme boundaries. Then the phonetic transcription (in SAMPA code) was aligned to the segmented phonemes (cf. Fig. 2).

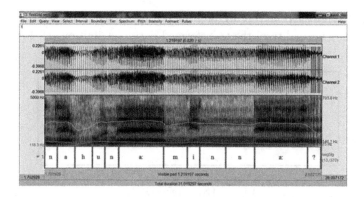

Fig. 2. Example of a sample segmentation and labeling

The obtained data had been stored in Praat Textgrid files which served afterwards to extract the prosodic parameters.

3.4 Prosodic Parameters Extraction

Parametric speech processing is based on spectral and prosodic parameters. Whereas spectral parameters, like MFCC, PLP, LPC ...etc. are mainly used in speech recognition, the prosodic parameters are necessary for speech synthesis, which is our main goal. The main prosodic parameter is the fundamental frequency (or pitch), which carries the intonation, whereas the parameters of duration and intensity express the timing and the energy of speech. The extraction of these parameters was done using the generated textgrid files and other functions of PRAAT tool. The duration is determined as the temporal difference between the phonemes boundaries, whereas F0 and intensity values are calculated at each 10 ms frame. Finally, the segmented phonemes and syllables were stored with these extracted parameters in a dedicated database, developed with mySQL.

4 Statisitcal Analysis

We remind that the main goal of this study is to determine the statistical distributions of the prosodic parameters, and the required transformations to normalize them, in order to decide which kind of speech to use in the next step of our Arabic text-to-speech project. Therefore a statistical analysis was conducted using Matlab toolboxes in order to determine the distribution of each prosodic parameter (F0, duration and intensity) at each level, to assess their normality and to look for the normalizing transformation in case they are not Gaussian.

4.1 Experimental Methodology

To achieve this task, an experimental plan was set: First, a normality test is done, then if the distribution is not Gaussian, the right kind of distribution is determined. The third step is to look for the normalizing transformation, and finally test the normality of the transformed data distribution (cf. Fig. 3).

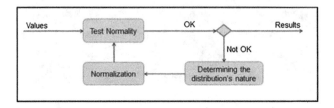

Fig. 3. Experimental plan

The first step, i.e. the normality test, is achieved by several techniques which assess the compatibility of an empirical distribution with the normal distribution. In this work, we relied on three tests [6]:

- The Kolmogorov-Smirnov test
- The Lilliefors test
- The Jarque-Bera test

If one of these test gives a normal distribution, the process is terminated, otherwise the loop continues. Data normalization consists in seeking the transformation through which the data follow a Gaussian distribution. Actually, the type of the normalizing transformation depends on the determined distribution. Therefore several transforms are used, mainly:

- Logarithmic transform: $Y = \log(x)$ (1)
- Square transform: $Y = x^2$ (2)
- Root square transform: $Y = \sqrt{x}$ (3)
- Arcsin transform: $Y = arcsin(\sqrt{x})$ (4)
- Exponential transform: $Y = e^x$ (5)
- Inverse transform: $Y = \frac{1}{x}$ (6)
- Sigmoid transform: $Y = \frac{1}{(1+exp(-x))}$ (7)
- Box-Cox transform: (8)
    - if $\lambda = 0$, then $Y(\lambda) = log(X)$
    - otherwise $Y(\lambda) = \frac{x^{\lambda-1}}{\lambda}$

4.2   Results

First, these tests were conducted on the prosodic parameters of phonemes and syllables of the spontaneous speech database. However, these sets are too broad to give homogeneous data and then normal distributions. Therefore, we chose to apply the experimental plan to every subset of phonemes (short and long vowels, semi-vowels, fricatives, nasals . . . etc.) and syllables (open/closed and short/long). For more details, an example for F0 distribution of closed syllables subset is mentioned hereafter.

F0 Distribution for Closed Syllables. We are considering here the samples of closed syllables stored in the spontaneous speech database.

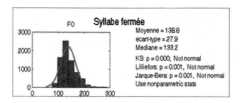

Fig. 4. Normality test for F0 closed syllables data

The first step is to check the normality of the distribution with a normality test. The results show that the collected values of F0for this type of syllables do not follow a normal distribution, whatever the type of the normality test (Kolmogorov-Smirnov, Lilliefors or Jarque-Bera) (cf. Fig. 4).

Determination of the Distribution Law. Since this distribution is not Gaussian, we move to the next stage, i.e. the identification of the nearest fitting distribution.

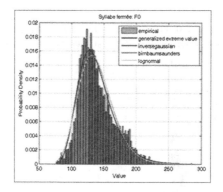

Fig. 5. Determination of the F0 distribution for closed syllables

The results show that the nearest fitting law is the "generalized extreme value" (GEV), also known as the law of Fisher-Tippet [8], which probability density is expressed in (9) (cf. Fig. 5).

$$f(x; \mu, \sigma, \xi) = \frac{1}{\sigma}\left[1 + \xi\left(\frac{x - \mu}{\sigma}\right)\right]^{\left(\frac{-1}{\xi}\right)-1} \tag{9}$$

where μ is the position parameter, σ the dispersion parameter and ξ is the shape parameter called index of extreme values.

Normalizing Transform. The results show that (a) for the distribution of duration, it is more normal for spontaneous speech than for prepared speech. However, it is possible to normalize it in all cases. (b) For the F0 distribution, it is more normal for the prepared speech. (c) The intensity distribution is not normal for any type of speech and any type of segments; moreover, it is not always possible to normalize it through any transformation (Fig. 6).

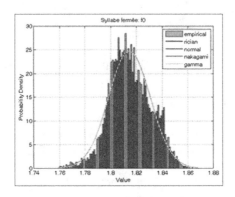

Fig. 6. Determination of the F0 distribution post-transformation

The BOXCOX transform is then applied for data normalization. Among the normality tests, the Jarque-Bera test shows that the transformed data are normally distributed (cf. Fig. 7).

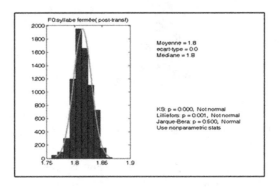

Fig. 7. Normality test post-transformation

4.3   Comparison with the Prepared Speech Database

A comparative survey was conducted for the distributions of prosodic parameters extracted, first from the spontaneous speech and secondly from the prepared speech database, which was processed in a previous work. The yielding results are mentioned in Tables 3, 4 and 5. As interpretation, the duration is the most

Table 3. Normality tests and normalizing transformations of duration

Type of segment	Spontaneous corpus distribution	Transformation	Prepared corpus distribution	Transformation
Open syllable	Log-logistic	Undefined	Gamma	Log
Closed syllable	Inverse Gaussian	Log	Log-logistic	Sigmoid
Short vowel	GEV	Square	GEV	log
Long vowel	GEV	Undefined	GEV	Log
Semi-vowel	Normal		Log-logistic	Log
Fricative consonant	Normal		Beta	Boxcox
Nasal consonant	Normal		Birnbaum Saunders	Log
Stop consonant	Nakagami	Boxcox	Nakagami	Boxcox

Table 4. Normality tests and normalizing transformations of F0

Type of segment	Spontaneous corpus distribution	Transformation	Prepared corpus distribution	Transformation
Open syllable	Log-logistic	Undefined	Weibull	Undefined
Closed syllable	GEV	Boxcox	Weibull	Undefined
Short vowel	GEV	Boxcox	Normal	
Long vowel	Gamma	Undefined	Normal	
Semi-vowel	Log-normal	Log	Normal	
Fricative consonant	Log-logistic	Boxcox	GEV	Boxcox
Nasal consonant	GEV	Undefined	GEV	Undefined
Stop consonant	Log-logistique	Boxcox	GEV	Boxcox

Table 5. Normality tests and normalizing transformations of intensity

Type of segment	Spontaneous corpus distribution	Transformation	Prepared corpus distribution	Transformation
Open syllable	Extreme value	Undefined	Extreme value	Boxcox
Closed syllable	Extreme value	Undefined	Extreme value	Boxcox
Short vowel	Weibull	Undefined	Extreme value	Undefined
Long vowel	Weibull	Boxcox	Extreme value	Undefined
Semi-vowel	Extreme value	Boxcox	Extreme value	Boxcox
Fricative consonant	GEV	Undefined	GEV	Undefined
Nasal consonant	Weibull	Boxcox	GEV	Undefined
Stop consonant	Extreme value	Undefined	GEV	Undefined

stable prosodic parameter, since it depends mainly from the language. Therefore it does not depend too much on the type of speech, whether spontaneous or prepared (read). Actually, the duration distributions are more normal for spontaneous speech than for read speech, especially for consonants. However, looking to Table 3, all read-speech durations can be normalized, mainly using the log transform, whereas it is impossible to normalize some spontaneous-speech durations with any transform, like for long vowels and open syllables (which themselves end by a vowel). This can be understood since Arabic pronunciation for long vowels follows many rules, where the length of the vowel is simple, double or triple according to its position in the syllable and the word. Then, in spontaneous speech these rules are not always well followed, whereas they are strictly respected in read speech. For F0 (cf. Table 4), it is more stable for prepared speech, since F0 expresses intonation. Then for a read speech, it is expected to have more control on pronunciation and then more normality in F0 distribution. Actually, the F0 distribution shows that F0 values look more normal for the prepared speech, especially for short, long and semi-vowels. This can be expected since all these types of segments are voiced. Also, most consonant segments (fricatives, nasals and stop consonants) follow a generalized extreme-value distribution, which would tend to be normal if the corpus had a bigger size. Finally, intensity is the most random prosodic parameter, especially for the spontaneous speech corpus (cf. Table 5). Since intensity describes the energy of speech, it depends, not only from the speech, but also from the speaker himself. However for the read speech, the intensity distribution follows an extreme-value law for short, long and semi-vowels and a generalized extreme-value law for all types of consonants, which shows more constancy in this kind of speech.

More generally, the spontaneous speech corpus gives more natural speech, which appears in the normal/normalizable durations, whereas the prepared speech offers more intelligibility, as can be guessed through the normal/normalizable F0, and more control in energy. Therefore, both kinds of corpus can be used to establish a speech database, which could be useful to synthesize natural and intelligible speech.

5   Conclusion

The analysis, annotation and labeling of a spontaneous Arabic speech was achieved in this work. Hence, a spontaneous Arabic speech corpus was collected, and analyzed to extract the prosodic parameters, i.e. duration, F0 and intensity, at different levels, i.e. the phoneme and the syllable. Then a statistical survey was conducted (a) to check the normality of the parameters distributions, (b) to determine the actual distributions and (c) to normalize them. Actually, normally distributed prosodic parameters may give better prediction models, especially when using statistical learning tools.

The results were compared to those of a prepared Arabic speech database, studied previously. This comparison shows that for duration, spontaneous speech gives more normal distributions, whereas for F0, the prepared speech gives better results. Then we can conclude that spontaneous speech offers more naturalness, whereas prepared speech ensures more intelligibility and control. Therefore, a mixed corpus containing spontaneous and prepared speech could be designed. Furthermore, an important finding, which is the normalizing transformations, could be used for future works aiming to model prosodic parameters using statistical learning for Arabic speech synthesis.

References

1. Abdel-Hamid, O., Abdou, S.M., Rashwan, M.: Improving Arabic HMM based speech synthesis quality. In: INTERSPEECH (2006)
2. Al-Ani, S.: Arabic phonology: an acoustical and a physilogical investigation. Walter de Gruyter (1970)
3. Boersma, P., Weenink, D.: Praat: doing phonetics by computer (2010)
4. Boudraa, M., Boudraa, B., Guerin, B.: Elaboration d'une base de données arabe phonétiquement équilibrée. In: Actes du colloque Langue Arabe et Technologies Informatiques Avancées, pp. 171–187 (1993)
5. Campbell, W.N.: Predicting segmental durations for accommodation within a syllable-level timing framework. In: 3rd European Conference on Speech Communication and Technology (1993)
6. Ghasemi, A., Zahediasl, S.: Normality tests for statistical analysis: a guide for non-statisticians. Int. J. Endocrinol. Metabol. $10(2)$, 486–489 (2012)
7. Ladd, D.R.: Intonational Phonology. Cambrige University Press, Cambridge (1986)
8. Markose, S., Alentorn, A.: The Generalized extreme value distribution and extreme economic value at risk (EE-VaR). In: Kontoghiorghes, E., Rustem, B., Winker, P. (eds.) Computational Methods in Financial Engineering. Springer, Berlin (2008)
9. Mixdorff, H., Jokisch, O.: An integrated approach to modeling German prosody. Int. J. Speech Technol. $6(1)$, 45–55 (2003)
10. Mnasri, Z., Boukadida, F., Ellouze, N.: Design and development of a prosody generator for Arabic TTS systems. Int. J. Comput. Appl. $12(1)$, 24–31 (2010)
11. Vainio, M., et al.: Artificial neural networks based prosody models for Finnish text-to-speech synthesis. Ph.D. thesis, Helsinky University of Technology (2001)
12. Van Santen, J.: Assignement of segmental duration in text-to-speech synthesis. Comput. Speech Lang. $8(2)$, 95–128 (1994)
13. Zen, H., Nose, T., Yamagishi, J., Sako, S., Masuko, T., Black, A.W., Tokuda, K.: The HMM-based speech synthesis system (HTS) version 2.0. In: SSW, pp. 294–299. Citeseer (2007)

Combining Syntactic and Acoustic Features for Prosodic Boundary Detection in Russian

Daniil Kocharov[1], Tatiana Kachkovskaia[1(✉)], Aliya Mirzagitova[2], and Pavel Skrelin[1]

[1] Department of Phonetics, St. Petersburg State University,
7/9 Universitetskaya nab., St. Petersburg 199034, Russia
{kocharov,kachkovskaia,skrelin}@phonetics.pu.ru
[2] Department of Mathematical Linguistics, St. Petersburg State University,
7/9 Universitetskaya nab., St. Petersburg 199034, Russia
amirzagitova@gmail.com

Abstract. This paper presents a two-step method of automatic prosodic boundary detection using both textual and acoustic features. Firstly, we predict possible boundary positions using textual features; secondly, we detect the actual boundaries at the predicted positions using acoustic features. For evaluation of the algorithms we use a 26-h subcorpus of CORPRES, a prosodically annotated corpus of Russian read speech. We have also conducted two independent experiments using acoustic features and textual features separately. Acoustic features alone enable to achieve the F_1 measure of 0.85, precision of 0.94, recall of 0.78. Textual features alone work with the F_1 measure of 0.84, precision of 0.84, recall of 0.83. The proposed two-step approach combining the two groups of features yields the efficiency of 0.90, recall of 0.85 and precision of 0.99. It preserves the high recall provided by textual information and the high precision achieved using acoustic information. This is the best published result for Russian.

Keywords: Speech prosody · Prosodic boundary · Syntactic parsing · Sentence segmentation · Automatic boundary detection · Statistical analysis

1 Introduction

Segmentation of speech and text into prosodic units is considered one of the key issues in speech technologies. Predicting prosodic boundaries from text using textual parameters—such as punctuation, constituent analysis, parts of speech etc.—is a crucial step in text-to-speech synthesis. Detecting prosodic boundaries in audio data—using acoustic features, such as pauses, fundamental frequency changes etc.—is an essential task in speech recognition.

For the purpose of creating new large speech corpora, combining these two tasks may appear extremely useful. Much of the audio data, which could be used for these corpora, are also stored in textual form: e.g., audiobooks, interviews etc.

© Springer International Publishing AG 2016
P. Král and C. Martín-Vide (Eds.): SLSP 2016, LNAI 9918, pp. 68–79, 2016.
DOI: 10.1007/978-3-319-45925-7_6

In such cases prosodic labeling of speech may rely on both syntactic and prosodic information.

So far our research team has been working on two separate tasks in parallel, using the same speech corpus:

- using textual data for prosodic boundary prediction;
- using acoustic features for prosodic boundary detection.

The present research is aimed at combining these two groups of cues in one system capable of predicting prosodic boundaries in speech. The paper includes descriptions of syntactic and acoustic components separately and in combination.

In real speech syntactic and prosodic boundaries do not always coincide. Thus, speakers often split syntactic constituents into two or more parts—due to pragmatic reasons, or if the whole phrase is too long. However, we assume that there are such word junctures where a prosodic boundary is highly improbable—e.g., between a preposition and its dependent noun. Based on this assumption, the syntactic component is trained to predict phrase boundaries with recall close to 100 %; as a result, the text is split into short phrases—mostly 1 or 2 words long. At the next step, these phrase boundaries are used as input to the prosodic component: it chooses among only those word junctures where a syntactic boundary is possible.

2   Material

The experiments were carried out on CORPRES—Corpus of Professionally Read Speech—developed at the Department of Phonetics, St. Petersburg State University [19]. The corpus contains recordings of various texts read by eight speakers.

The total duration of the recorded material is 60 h; all of it is prosodically annotated. The prosodic annotation is stored in text files and was performed as follows:

- each utterance is divided into intonational phrases (tone units);
- for each intonational phrase, the lexical word carrying nuclear accent is marked and the melodic type is assigned;
- words carrying additional prosodic prominence are marked with a special symbol.

The prosodic annotation was performed by expert phoneticians using perceptual and acoustic data.

Half of the recorded material—30 h—is segmented into intonational phrases (tone units), lexical words, and sounds. Prosodic tier was generated based on prosodic information from the text files described above. Phonetic tier was added manually based on perceptual and acoustic analysis (using spectral data if necessary). Stress was marked on the phonetic tier based on actual pronunciation.

For the experiments based on textual data, we have chosen three texts recorded from at least four speakers each[1]:

---

[1] Texts A, B and C comprise 75 % of all the recordings.

– text A: a fiction narrative of rather descriptive nature containing long sentences and very little direct speech (19,110 words);
– text B: an action-oriented fiction narrative resembling conversational speech (16,769 words);
– text C: a play containing a high number of conversational remarks and emotionally expressive dialogues and monologues (21,876 words).

Each of these texts was recorded from several speakers (4–8), which enables us to take into account the prosodic boundary placement across speakers. The texts containing prosodic annotation were automatically aligned between the speakers; then for each word juncture we calculated the number of speakers who placed a prosodic boundary there. A boundary between two adjacent words was considered possible if it was observed for least two speakers, since a boundary produced by only one speaker may be occasional, and evidence from two and more speakers we may reflect a tendency. Thus we obtained all possible prosodic boundaries in texts A, B, and C. This textual material corresponds to around 45 h of read speech.

The segmented part of the corpus was used for the experiments based on acoustic data. Along with texts A, B, and C, this part also included newspaper articles on IT (D) and on politics and economics (E); they comprise around 12 % of the segmented material. The total duration of this material is around 30 h of speech.

Using the data on stress from the phonetic tier, the corpus was segmented into prosodic words[2]. Since a boundary is placed between two adjacent prosodic words, we analysed prosodic boundary placement for each such pair.

In order to test the combination of syntactic and acoustic cues for prosodic boundary detection, we used the overlap of the two subsets of the corpus described above: the segmented part of texts A, B, and C. The total duration of the overlap is 26 h of speech.

3   Method

The most common approaches to prosodic boundary prediction are rule-based methods [1,14,17]; data-driven probabilistic methods: N-grams [21], probabilistic rules [5], weighted regular tree transducer [22]; machine learning: memory-based learning [3], HMM [11,16], CART and random forests [9], neural networks and support vector machines [7].

The most promising results on phrase boundary detection in Russian texts were obtained when using segmentation of text into chunks within which prosodic boundaries are impossible [9,12]. The efficiency reported in [9] estimated by F1 measure is 0.76. The main disadvantage of these systems is that the construction of chunking rules was done manually by experts, which is costly and time-consuming.

---

[2] We use the term "prosodic word" in its traditional sense for a content word and its clitics, which lose their lexical stress and form one rhythmic unit with the adjacent stressed word.

In this paper we propose a fully automatic approach for prosodic boundary detection based on both syntactic and acoustic features. The procedure consists of two steps. As a first step, we obtain text-based predictions for prosodic boundaries; text chunking is performed based on the syntax tree produced by a dependency parser. The next step is extracting prosodic data: a set of acoustic features calculated from speech signal. The extracted acoustic features are used to classify the word junctures predicted at the first step, i.e. after text processing.

In this section we describe textual and acoustic features used for phrase boundary detection, and statistical classifiers used for the task.

3.1   Textual Features

A wide range of textual features has been used in different systems for prosodic boundary prediction: punctuation and capitalization [9], estimation of phrase length [21], word class (content word vs. function word) [21], n-grams of part-of-speech tags [15,21], shallow constituent-based syntactic methods [1,9,12], deep syntactic tree features [5,6,16,22].

The set of features used in our experiments is described above.

Punctuation. Punctuation serves to split written text into meaningful pieces in a similar way as prosody does with speech. Since there is a strong correspondence between punctuation marks and prosodic boundaries, we use punctuation marks within a word juncture as a prosodic boundary marker. It should be noted, though, that in a number of cases punctuation marks do not require prosodic boundaries [4], e.g. commas used to separate series of terms or set off introductory elements, full stops at ends of abbreviations.

Phrase Length. Apart from the informational structure, speech segmentation into prosodic units is also regulated by physiological mechanisms [23]. This is one of the reasons why speakers tend to split long syntactic phrases into shorter ones. Thus, in Russian read speech the average length of an intonational phrase is 7.5 syllables, with a standard deviation of around 4 syllables [24].

Therefore, phrase length should be taken into account when predicting prosodic boundaries. As estimates of phrase length, we are using the number of words and the number of syllables between the current juncture and the previous boundary; the number of syllables is calculated as the number of vowel letters.

Part-of-Speech Tags. It has been shown that part-of-speech tags are good predictors of prosodic boundaries [21]. Thus, in Russian certain parts of speech are never followed by a prosodic boundary, e.g. prepositions; on the other hand, some parts of speech tend to appear at ends of prosodic units, e.g. verbs. In our experiments, we consider part-of-speech tags an important feature for juncture classification.

Syntactic Features. In general, linguists agree that there exists a correspondence between syntactic and prosodic structure of utterances [1,23]. Deep and shallow syntactic features have been successfully used in previous studies for phrase boundary prediction.

It has been shown that prediction based on word junctures where a boundary is impossible works well for Russian [9,12]. This approach has been taken in our experiments.

We use three syntactic markers for word juncture classification: presence of syntactic phrase boundary; part of speech of the main word within the current syntactic phrase; type of the link between the main word within the current syntactic phrase and its parent.

Splitting of text into syntactic phrases requires text to be tagged and syntactically parsed. Since there are no available shallow parsers for Russian, as well as open syntactically tagged Russian corpora, we had to perform a full syntactic analysis of the data.

We used Solarix, a free open-source tool, for a full morphosyntactic analysis [10]. It has built-in lexicons and pre-trained models for processing both Russian and English languages. The parser accepts plain text as input, splits it into individual sentences and tokens, performs probabilistic part-of-speech tagging and builds a dependency tree structure by choosing the most probable syntactic link type for each word pair.

A set of rules was developed to map the obtained syntactic tree structures to the linear order of phrases. It should be noted that we did not intend to extract the complete adjacent groupings from sub-trees (with more than three constituents), because it was assumed that there would be boundaries inside them in real speech. The main rule was that phrase boundaries are placed at those junctures where syntactic discontinuity is observed. Long phrases without syntactic discontinuity were divided further using the following principle: a boundary was placed before a word if its parent stood next to it, but formed its own group with another word, e.g. [the development of algorithms for solving equations] → [the development of algorithms][for solving equations]. Another rule concerns prepositions. In cases of syntactic discontinuity between the preposition and the following word, the preposition was automatically assigned to the following group: e.g. [stands on][a white table] → [stands][on a white table]. Thus, we obtained a fine-grained segmentation of sentences into sequences of phrases.

3.2   Acoustic Features

A number of acoustic features have been used for prosodic boundary detection. Common durational features include pause duration and various estimates of pre-boundary lengthening. These cues show high efficiency, and some systems only durational features [25,26]. Features based on fundamental frequency include F_0 range, F_0 slope, mean F_0 and others [7,17]. Intensity, amplitude or energy are considered weaker cues [20], but are also used in some systems [7,11].

The beginning of a new utterance or intonational phrase (IP) is often marked by a reset of acoustic features: fundamental frequency, tempo, amplitude, which can be explained by underlying mechanisms of speech production. In general, a new utterance or IP often begins with higher F_0, faster tempo and higher amplitude [23].

A part of acoustic features chosen for prosodic boundary prediction are based on this phenomenon. For each pair of adjacent prosodic words in the material, we calculated the difference (reset) in F_0, tempo and amplitude between them. Besides these, we used several standard features: pause duration between the two prosodic words, and F_0 range values for the first and the second prosodic word in the pair.

Pause. Silent pauses are known to be the most reliable cue for boundary detection. However, by no means all of the boundaries are marked in this way—it is rather common to observe a sequence of intonational phrases produced without pauses.

Are another important cue—filled pauses, or hesitations—is not used in the present study, since they are very rare in our material.

Segmental Duration. It is known that in many languages including Russian, speakers tend to slow down towards the end of the utterance or intonational phrase; this phenomenon is often called pre-boundary lengthening. We estimate this slowdown by calculating tempo reset: for each pair of adjacent prosodic words, it is the difference between their stressed vowel's normalized duration values.

For phone duration normalization, the following formula was used, which allows to compensate for the average duration of the segment and its standard deviation:

$$\tilde{d}(i) = \frac{d(i) - \mu_p}{\sigma_p}$$

where $\tilde{d}(i)$ is the normalized duration of segment i, $d(i)$ is its absolute duration, and μ_p and σ_p are the mean and standard deviation of the duration of the corresponding phone p. The mean and standard deviation are calculated over the whole corpus for each speaker separately.

The major reason to measure tempo reset using the stressed vowel's duration is because in Russian the stressed vowels are the main carriers of pre-boundary lengthening [8].

Fundamental Frequency. The typical F_0 contour of a simple sentence is a sequence of rise-falls corresponding to prosodic words, where the phenomenon of declination is observed [23]. Mainly, F_0 declination implies the following: (1) F_0 maximum is located within the first word of the sentence; (2) F_0 range decreases towards the end of the sentence.

Another phenomenon concerns the F_0 contour within the nucleus. In Russian, nuclear stress, where a major F_0 change is often observed, is usually located on the last prosodic word within the intonational phrase.

Based on these tendencies, we are using three features:

- F_0 reset—for each pair of adjacent prosodic words, it is the difference between their stressed syllable's F_0 values;
- F_0 range (in semitones) on the stressed syllable of the first prosodic word in each pair of adjacent words—to detect the F_0 change on the nucleus;
- F_0 range (in semitones) on the stressed syllable of the second prosodic word in each pair of adjacent words—to detect the F_0 change on the first prosodic word in the IP.

Amplitude. The declination of F_0 contour if often matched with the declination of intensity. So far, we estimate this reset using amplitude values: for each pair of adjacent prosodic words, it is the difference between their stressed vowel's mean amplitude values.

3.3   Statistical Analysis

Conditional Random Fields (CRFs). In terms of syntax, we considered the task of prosodic phrase segmentation as being similar to shallow parsing, or chunking, that is dividing a sentence into non-overlapping syntactic units. In its turn, it is often handled as a sequential classification problem, an approach to which is to employ the conditional random fields (CRFs) model [18]. We evaluated whether it is also applicable to predict and assign prosodic phrase boundaries in a text. To train the CRFs model, we used the MALLET implementation [13].

Random Forests (RFs). The classification of word junctures based on prosodic information was performed by means of Random Forests classifier [2]. The set of word junctures was considered a homogeneous set of individual units, but not an organized sequence of units.

Evaluation Criteria. We use common measures to assess our procedures for comparability with other systems and researches, i.e. precision (Pr), recall (R) and F_1 measure. Precision is a number of correctly predicted boundaries relative to total number of predicted boundaries. Recall is a number of correctly predicted boundaries relative to total number of boundaries in corpus. F_1 measure is calculated by the following formula:

$$F_1 = 2 \cdot \frac{Pr \cdot R}{Pr + R} \tag{1}$$

4   Results

This section contains a description of experimental results obtained when using only textual information, when using only acoustic information, and when using combination of textual and acoustic information. They are presented successively.

It is worth noting that 37.3 % of all the words in the material are preceded by a prosodic boundary. Therefore, a set of features yielding a 100 % recall and precision of 37.3 % is unusable, since it predicts a boundary after each word; this is used as a rough baseline to determine the efficiency of an experimental design.

4.1   Syntactic Features

For training and testing of CRFs we used different combinations of the texts, see Table 1. The texts are different from the syntactic point of view, as it was described in Sect. 3. The features used for classification are presented in Sect. 3.1. The values of precision, recall and F_1 measure are reported in Table 1.

Table 1. Perfomance of CRFs for different configurations

Training		Testing		Precision	Recall	F_1
Text	Sentences	Text	Sentences			
A and B	**3876**	**C**	**6136**	**0.90**	**0.89**	**0.89**
A and C	7853	B	2159	0.81	0.81	0.81
B and C	8295	A	1717	0.77	0.77	0.77
Mean				0.84	0.83	0.84

All the experimental designs presented here show better efficiency compared with the results published previously for Russian, see [9].

We decided to train the classifier on different texts in order to find out whether the syntactic organization of the text influences the efficiency of prosodic boundary prediction. The results show that texts A and B, being syntactically richer, predict boundaries better than text C, containing simple, colloquial speech. This means that syntactic models for prosodic boundary prediction show better results when trained on prosodically rich texts, even if we intend to use them for conversational speech.

4.2   Acoustic Features

Performance of Individual Acoustic Features. Table 2 provides performance data for each of the acoustic features presented in Sect. 3.2. F_1 measure given in this table was calculated as the maximum F_1 measure over all possible threshold values for the given feature.

Overall, these data confirm that durational cues are the most effective [20], along with the F_0 range for the first prosodic word in a pair of two adjacent prosodic words. Amplitude and fundamental frequency resets show lower performance, but may still matter in combination with other features.

The last feature, F_0 range within the second word's stressed syllable, proves to be inefficient, since it provides the same results as the baseline design predicting a boundary after each word.

Table 2. Performance of individual acoustic features

Feature	Precision	Recall	F_1
Pause duration	**0.98**	**0.75**	**0.85**
Tempo reset	0.55	0.70	0.61
F_0 range (first word)	0.50	0.77	0.60
Amplitude reset	0.50	0.70	0.58
F_0 reset	0.45	0.53	0.48
F_0 range (second word)	0.37	1.00	0.54

Performance of Combination of Acoustic Features. Table 3 provides performance data for the combination of all acoustic features using Random Forest classifier. The results show that acoustic features give rather high performance; the order of features as to their significance for classification is the same as in Table 2.

To demonstrate the significance of pauses as a boundary cue, we also calculated the performance of all the keys combined without data on pausation (see Table 3). Note that pausation does not affect the recall, but seriously affects the precision. High precision is explained by the fact that pauses almost always occur at prosodic boundaries. At the same time, a boundary is not always accompanied by a pause, and this explains the unchanged recall.

Table 3. Performance of combinations of acoustic features

Features	Method	Precision	Recall	F_1
Prosodic, with pauses	Random forest	0.94	0.78	0.85
Prosodic, without pauses	Random forest	0.68	0.78	0.73

4.3   Combination of Syntactic and Acoustic Features

The general design of classification procedure was described in Sect. 3. We use a two-step procedure.

At the first step, we use syntactic phrase boundaries as predictors of prosodic boundaries. Table 4 demonstrates their efficiency. It can be seen that it ensures high recall while having F_1 measure above 0.69. We assume that the false negative errors were due to the tokenization problems within the syntactic parser.

Table 4. Performance of prosodic boundary prediction on different texts

Text	Precision	Recall	F_1
A	0.58	0.97	0.73
B	0.53	0.97	0.69
C	0.64	0.97	0.77
Mean	0.61	0.97	0.75

If we change the syntactic phrasing rules so that the recall is as high as 0.99, the precision drops to 0.40. As already discussed in Sect. 4.2, such result would be misleading as the potential boundary would be predicted in almost every word juncture. So we decided to leave imperfect recall moving 3 % of boundaries to word junctures without boundaries. This shift of performance was accounted for in the evaluation of overall performance presented in Table 5.

At the second step, we used Random Forests classifier with all acoustic features for classification of word junctures predicted at the first step.

In combination, syntactic and acoustic features perform better than separately. The recall is as high as when using syntax alone, and the precision is as high as when using prosodic information only. The efficiency measures are also better: we obtain F_1 measure of 0.90, which is the highest reported result for Russian and among the highest for other languages.

Table 5. Comparison of the results obtained in all the experiments

Experiment	Precision	Recall	F_1
CRFs with syntactic features	0.84	0.83	0.84
RFs with acoustic features	0.94	0.78	0.85
Two-step procedure with syntactic and acoustic features	**0.99**	**0.85**	**0.90**

5 Discussion

In these experiments, performed on read speech, a combination of syntactic and acoustic features for automatic prosodic boundary detection has shown high efficiency. The role of syntactic component was rather high, but here it was due to the material: read speech is syntactically well-organized, and therefore phrasing is predictable.

Our next step is to develop this system for predicting prosodic boundaries in spontaneous speech. Spontaneous speech differs from read and prepared speech in both syntax and prosody. Real-time speech planning often leads to complications in syntactic parsing. Thus, in spontaneous speech we often deal with interrupted or unfinished phrases, mispronunciations of word endings carrying principal grammatical information, non-linear word order where closely-linked constituents are split up by other words etc.

Therefore, for spontaneous speech we expect lower performance of syntactic features. For read speech, we used syntax first and then added acoustic data. For spontaneous speech we might need to begin with acoustics and only then pass on to syntax. This will also require the use of more acoustic features, such as global temporal, dynamic, and melodic features; hesitations and elongations; types of non-speech events etc. The effectiveness of pause duration, which is very high in read speech, may decrease, since in spontaneous speech the phonation is interrupted not only to signal a boundary, but also to take a breath or think of what is about to be said next.

So far, our system has taken into account the acoustic data calculated for each individual word juncture. By not observing speech signal as a sequence of words, we may lose some important information, such as global F_0 changes. A finite-state automaton may be a good solution, and this will require some changes in the set of acoustic features used for the analysis.

Acknowledgments. The research is supported by the Russian Science Foundation (research grant # 14-18-01352).

References

1. Bachenko, J., Fitzpatrick, E.: A computational grammar of discourse-neutral prosodic phrasing in English. Comput. Linguist. **16**(3), 155–170 (1990)
2. Breiman, L.: Random forests. Mach. Learn. **45**(1), 5–32 (2001)
3. Busser, B., Daelemans, W., van den Bosch, A.: Predicting phrase breaks with memory-based learning. In: Proceedings of the 4th ISCA Tutorial and Research Workshop on Speech Synthesis, pp. 29–34 (2001)
4. Chafe, W.: Punctuation and the prosody of written language. Writ. Commun. **5**(4), 395–426 (1988)
5. Hirschberg, J., Rambow, O.: Learning prosodic features using a tree representation. In: Proceedings of Eurospeech 2001, pp. 1175–1178 (2001)
6. Hoffmann, S.: A data-driven model for the generation of prosody from syntactic sentence structures. Ph.D. thesis, ETH-Zürich, Zürich (2014)
7. Jeon, J.H., Liu, Y.: Semi-supervised learning for automatic prosodic event detection using co-training algorithm. In: ACL 2009, Stroudsburg, PA, USA, vol. 2, pp. 540–548. Association for Computational Linguistics (2009)
8. Kachkovskaia, T.: The influence of boundary depth on phrase-final lengthening in Russian. In: Dediu, A.-H., et al. (eds.) SLSP 2015. LNCS, vol. 9449, pp. 135–142. Springer, Heidelberg (2015). doi:10.1007/978-3-319-25789-1_13
9. Khomitsevich, O., Chistikov, P., Zakharov, D.: Using random forests for prosodic break prediction based on automatic speech labeling. In: Ronzhin, A., Potapova, R., Delic, V. (eds.) SPECOM 2014. LNCS, vol. 8773, pp. 467–474. Springer, Heidelberg (2014)
10. Koziev, E.: Solarix (2016). http://www.solarix.ru

11. Liu, Y., Shriberg, E., Stolcke, A., Hillard, D., Ostendorf, M., Harper, M.: Enriching speech recognition with automatic detection of sentence boundaries and disfluencies. IEEE Trans. Audio Speech Lang. Process. **14**(5), 1526–1540 (2006)

12. Lobanov, B.: An algorithm of the text segmentation on syntactic syntagrams for TTS synthesis. In: Proceedings of Dialogue 2008 (2008)

13. McCallum, A.K.: MALLET: a machine learning for language toolkit (2002). http:// mallet.cs.umass.edu

14. Ostendorf, M., Veilleux, N.: A hierarchical stochastic model for automatic prediction of prosodic boundary location. Comput. Linguist. **20**(1), 27–54 (1994)

15. Read, I., Cox, S.: Using part-of-speech tags for predicting phrase breaks. In: Proceedings of Interspeech 2004, Jeju Island, Korea, pp. 741–744, October 2004

16. Read, I., Cox, S.: Stochastic and syntactic techniques for predicting phrase breaks. Comput. Speech Lang. **21**(3), 519–542 (2007)

17. Segal, N., Bartkova, K.: Prosodic structure representation for boundary detection in spontaneous French. In: Proceedings of ICPhS 2007, pp. 1197–1200 (2007)

18. Sha, F., Pereira, F.: Shallow parsing with conditional random fields. In: Proceedings of NAACL 2003, pp. 134–141 (2003)

19. Skrelin, P., Volskaya, N., Kocharov, D., Evgrafova, K., Glotova, O., Evdokimova, V.: CORPRES - corpus of Russian professionally read speech. In: Sojka, P., Horák, A., Kopeček, I., Pala, K. (eds.) TSD 2010. LNCS, vol. 6231, pp. 392–399. Springer, Heidelberg (2010)

20. Streeter, L.A.: Acoustic determinants of phrase boundary perception. J. Acoust. Soc. Am. **64**(6), 1582–1592 (1978)

21. Taylor, P., Black, A.W.: Assigning phrase breaks from part-of-speech sequences. Comput. Speech Lang. **12**(2), 99–117 (1998)

22. Tepperman, J., Nava, E.: Where hould pitch accents and phrase breaks go? A syntax tree transducer solution. In: Proceedings of Interspeech 2011, pp. 1353–1356 (2011)

23. Vaissire, J.: Language-independent prosodic features. In: Cutler, A., Ladd, D.R. (eds.) Prosody: Models and Measurements. Springer Series in Language and Communication, vol. 14, pp. 53–66. Springer, Heidelberg (1983)

24. Volskaya, N.: Prosodic features of Russian spontaneous and read aloud speech. In: de Silva, V., Ullakonoja, R. (eds.) Phonetics of Russian and Finnish, pp. 133–144. Peter Lang, Bern (2009)

25. Wightman, C.W., Ostendorf, M.: Automatic recognition of prosodic phrases. In: Proceedings of ICASSP 1991, vol. 1, pp. 321–324 (1991)

26. Yoon, T., Cole, J., Hasegawa-Johnson, M.: On the edge: acoustic cues to layered prosodic domains. In: Proceedings of ICPhS 2007, Saarbrcken, Germany, pp. 1264–1267 (2007)

Articulatory Gesture Rich Representation Learning of Phonological Units in Low Resource Settings

Brij Mohan Lal Srivastava$^{(\boxtimes)}$ and Manish Shrivastava

Language Technology Research Center,
International Institute of Information Technology, Hyderabad, India
brijmohanlal.s@research.iiit.ac.in, m.shrivastava@iiit.ac.in

Abstract. Recent literature presents evidence that both linguistic (phonemic) and non linguistic (speaker identity, emotional content) information resides at a lower dimensional manifold embedded richly inside the higher-dimensional spectral features like MFCC and PLP. Linguistic or phonetic units of speech can be broken down to a legal inventory of articulatory gestures shared across several phonemes based on their manner of articulation. We intend to discover a subspace rich in gestural information of speech and captures the invariance of similar gestures. In this paper, we investigate unsupervised techniques best suited for learning such a subspace. Main contribution of the paper is an approach to learn gesture-rich representation of speech automatically from data in completely unsupervised manner. This study compares the representations obtained through convolutional autoencoder (ConvAE) and standard unsupervised dimensionality reduction techniques such as manifold learning and Principal Component Analysis (PCA) through the task of phoneme classification. Manifold learning techniques such as Locally Linear Embedding (LLE), Isomap and Laplacian Eigenmaps are evaluated in this study. The representations which best separate different gestures are suitable for discovering subword units in case of low or zero resource speech conditions. Further, we evaluate the representation using Zero Resource Speech Challenge's ABX discriminability measure. Results indicate that representation obtained through ConvAE and Isomap out-perform baseline MFCC features in the task of phoneme classification as well as ABX measure and induce separation between sounds composed of different set of gestures. We further cluster the representations using Dirichlet Process Gaussian Mixture Model (DPGMM) to automatically learn the cluster distribution of data and show that these clusters correspond to groups of similar manner of articulation. DPGMM distribution is used as apriori to obtain correspondence terms for robust ConvAE training.

Keywords: Neural representation of speech and language · Manifold learning · Zero resource · Articulatory gestures

© Springer International Publishing AG 2016
P. Král and C. Martín-Vide (Eds.): SLSP 2016, LNAI 9918, pp. 80–95, 2016.
DOI: 10.1007/978-3-319-45925-7_7

1   Introduction

Speech recognition is highly successful in scenarios where labelled speech data is available. However this is scarce in case of languages which are under-studied or under-resourced. There is a need for a representation which can discriminate between different speech sounds based on the underlying linguistic structure hidden in speech signal. Speech is a combination of several articulatory gestures shared across phonetic units. Browman and Goldstein presented several studies theorizing articulatiry gestures as the atomic units which compose phonological structures [5–8]. For instance stop consonants /p/ and /b/ share the sudden release of pressure built up behind the lips at their end, whereas they differ during the initial phase of total closure when vocal cords vibrate in case of /b/ but not while articulating /p/. In case of languages for which we do not have a list of discrete phonetic units, these recurring sub-phonetic gestures can be discovered by categorizing similar patterns in a representation which successfully captures the perceptual invariance of their hidden linguistic structure.

The varying nature of speech signal is due to various speaker-dependent signatures. It also contains para-linguistic information such as gender, speaker identity, emotions, accents, etc. Regardless of these inherent variations, humans can naturally distinguish between different sounds and categorize similar sounds. As stated by Blumstein [4] and Greenberg [12], humans recognize speech sounds by perceiving the gross characteristics of signal spectrum in frequency domain. This motivates the problem of identifying the neural mechanism responsible for identifying phonemic units present in speech based on acoustic variations produced by independent movement of articulators. Being able to figure out this phenomenon will help us move towards the bottom-up approach to create a discrete phonetic inventory and develop better speech recognition systems for under-resourced languages. According to Ostendorf, [20] such linguistically or acoustically motivated sub-word units are better suited for the task of continuous speech recognition as opposed to the traditional 'beads on a string' model where a word is assumed to be composed of a sequence of discrete phone segments. Lee and Siniscalchi [16] describes the steady progress of speech understanding over the years, starting from the bottom-up knowledge driven approaches to top-down learning-based data-driven approaches. Lee also proposed a framework for automatic speech attribute transcription (ASAT) which attempts to mimic human speech recognition capabilities using speech event detection followed by integrating the knowledge of articulatory gestures (bottom-up knowledge) for verification and recognition. This work proposes an automated approach towards learning such sub-word units from limited amounts of data.

The problem of learning discriminative representations for speech sounds is closely related to the problem of Zero Resource Speech Challenge (ZRSC) [1]. Therefore we evaluate the dimensionality reduction methods using ZRSC's ABX discrimination measure. This challenge focuses on discovering fundamental linguistic units from raw speech without supervision as learnt by human infants. This problem can also be posed as an analogous problem of discovering the neural mechanism of human brain responsible for learning and understanding linguistic information with such high efficiency.

Several recent studies [10, 25–27] focus on discovering the geometric structures in data, which is based on the premise of manifold hypothesis. Non-linear manifolds have also been explored to discover para-linguistic information like emotions [28] in speech. According to manifold hypothesis real-world data such as natural stimuli for speech tends to concentrate near a low-dimensional manifold embedded in high-dimensional data. Therefore we compare several unsupervised dimensionality reduction techniques using phoneme classification task. Though phoneme classification is just a preliminary benchmark, it enables us to ascertain the feature transformation technique which induces high separability of speech sounds. Several architectures of standard and convolutional autoencoders were trained to reconstruct the input features i.e. MFCCs extracted from raw speech. We also compared 7 manifold learning techniques namely 4 variants of Locally Linear Embeddings (LLE) along with Isometric Mapping (IsoMap) [24], Laplacian Eigenmaps [3] and PCA. Manifold learning tries to characterize a low-dimensional region in the input space with high data density. It usually involves construction of a neighbourhood graph using pairwise similarity between samples followed by partial eigenvalue decomposition to obtain low-dimensional embedding. In the next section we will provide a brief explanantion of each method considered in this study. Section 3 describes the experimental setup including dataset, training and evaluation criteria. In Sect. 4 we discuss the results and observations over the obtained representations. Section 5 concludes the paper and mentions some of the future directions.

2   Models

These techniques try to learn representations directly from data with help of some very generic priors introduced at feature level such as Mel-frequency filter banks, linear prediction, etc. Autoencoders have been previously used in [14, 22] and [2] for representation learning of speech sounds in an unsupervised manner to address practical problems like Zero Resource Speech Challenge. Autoencoders are highly expressive since they can represent exponential amount of input configurations as function of its number of parameters. Moreover, the representations obtained from autoencoders does not simply capture the local neighbourhood of input regions but associates various regions of input space together based on shared set of parameters. Deep autoencoders help re-use the features since the number of path through which input is transformed grows exponentially. With each layer of depth, the abstraction of feature also increases. Convolutional autoencoders build this abstraction by application of max- or average-pooling after each convolution layer. This property can be used to make the representation invariant to local variability of speech signal caused by several factors and increase its predictive capability.

Speech is highly entangled with factors of dynamic variations. These explanatory factors, such as articulatory movements, often tend to change either independent of each other or as a consequence of other factors. Speech signal displays this variability and the goal of a neural mechanism should be to learn to decompose these factors out of raw signal and create explanatory representations. For learning features suitable for keyword spotting we need representations which can overcome the variations of speaker and capture linguistically or gesturally invariant set of features directly from speech. We investigate the following low-dimensional transformations.

2.1 Autoencoder

Linear manifold learning methods such as PCA fail to capture abstract features which can be captured by deeper non-linear transformations such as autoencoders. Autoencoders learn a non-linear parametric mapping $f_\theta(x)$ from input space to a latent representation h which maps back to the input space using a decoder function $g_\theta(h)$. Once this mapping is learnt, hidden representation h can be efficiently computed and used as the feature vector. Stochastic gradient descent can be used to learn the set of parameters θ, which minimizes the reconstruction error given by $L(x, r)$.

$$h = f_\theta(x) \tag{1}$$

$$r = g_\theta(h) \tag{2}$$

$$L(x, r) = ||x - r||^2 \tag{3}$$

Here, f and g are encoder and decoder functions, which are nothing but affine transformation of their inputs followed by a non-linearity. We have experimented with various non-linearities including sigmoid, rectified linear units (ReLU), hyperbolic tangent (tanh) and parametric rectified linear units (PReLU) [13]. Tanh and PReLU perform comparatively better than other activation functions.

Parametric ReLU is a generalized form of ReLU where the coefficient of negative part of the function is not constant (or zero as in ReLU). It is a parameter which can be adaptively learnt. Formally, it can be defined as:

$$f(x) = \begin{cases} x, & \text{if } x > 0 \\ ax, & \text{if } x \leq 0 \end{cases}$$

Here x is the input to PReLU function and a is the learnable parameter which controls the slope of the negative part of the function. In case of ReLU, $a = 0$.

We chose tanh as the activation function for all our further experiments. Hence f and g are given by:

$$f_\theta(x) = tanh(Wx + b) \tag{4}$$

$$g_\theta(h) = tanh(W'h + d) \tag{5}$$

In a related study it has been shown that Adam [15] by Kingma et al. performs better than standard mini-batch SGD. We ascertain it by experimentation. Adam perform first-order gradient optimization based on adaptive estimates of lower-order moments of objective function. SGD selects a subset of training sample and compute the gradient using the mini-batch. This drastically increases the convergence rate. Adam tries to solve the quirks of stochastic gradient descent by devising a complex strategy to compute the direction and step size of each SGD iteration. It combined the advantages of AdaGrad and RMSProp in order to work well with sparse gradients in non-stationary settings. The central idea of Adam is to maintain exponential moving averages of gradient and its square. Following Algorithm 1 gives a brief idea about the weight update strategy of Adam:

Algorithm 1. Adam learning algorithm

1: **procedure** ADAM–WEIGHT–UPDATE
2: $M_0 = 0, R_0 = 0$ (Initialization)
3:     **for** t = 1 ... T **do**
4:         $M_t = \beta_1 M_{t-1} + (1 - \beta_1)\nabla L_t(W_{t-1}))$ (1st moment estimate)
5:         $R_t = \beta_2 R_{t-1} + (1 - \beta_2)\nabla L_t(W_{t-1})^2$ (2nd moment estimate)
6:         $\hat{M}_t = M_t/(1 - (\beta_1)^t)$ (1st moment bias correction)
7:         $\hat{R}_t = R_t/(1 - (\beta_2)^t)$ (2nd moment bias correction)
8:         $W_t = W_{t-1} - \alpha\dfrac{\hat{M}_t}{\sqrt{\hat{R}_t}+\epsilon}$
9:         Return W_t
10:     **end for**
11: **end procedure**

Here α is the learning rate which usually takes a positive value between 0.1 to 0.0001. $\beta_1 \in [0, 1)$, $\beta_2 \in [0, 1)$ is the 1st and 2nd moment decay rate respectively. $\epsilon > 0$ is the numerical term representing the threshold. M_t and R_t are the 1st and 2nd order moments of the gradient at time t.

2.2   Convolutional Autoencoder (convAE)

Learning to reconstruct the input from hidden representation is not enough to generalize the parameters to real-world input. In the denoising autoencoder architecture, the input is artificially corrupted by Gaussian noise and the network is forced to learn the clean version of input. Thus the parameters learn to undo the effect of noise and create a representation which reside in the high-density region of input space.

Convolutional architecture have been shown to perform state-of-the-art results in image classification tasks due to its superior feature learning technique. Figure 1 depicts a common architecture of convAE. In a conventional Convolutional Neural Network (CNN), a set of kernel matrices are initialized to learn different subsets of input characterizing the local receptive fields over the input. These fields are sweeped over the input to extract features which result in feature maps. Further these feature maps are subjected to local summarization by taking max or average of a window known as pooling. This mechanism aids to learn gross characteristics of input and removes sensitivity of representation in the direction of variance of input. Convolutional autoencoders have been recently shown to display impressive results to obtain better representation for object recognition tasks [17–19].

We combine denoising and convolutional architecture with autoencoders to learn robust and invariant representation of speech sounds. The following equation show the representation obtained by k-th feature map:

$$h^k = tanh(W^k * \tilde{x} + b^k) \tag{6}$$

where $*$ represents the convolution operation over speech feature such as MFCC, $\tilde{x}$ is the corrupted input feature, W^k are the weights of k-th kernel matrix and b^k is the bias term.

Input can be reconstructed using:

$$r = tanh(\sum_k h^k * W'^k + d) \tag{7}$$

The loss function to minimize is similar to Eq. 3.

DPGMM Clusters. DPGMM is a non-parametric bayesian mixture modeling technique which does not require explicit mention of number of components unlike traditional GMMs [11]. DPGMM assumes Dirichlet process as the prior distribution over infinite components representing the clusters. For practical purpose we assume a truncated maximum number of components with normally distributed means and identity covariance. We can choose the concentration parameter, α of Dirichlet process which intuitively represents the spread of data. Experiments have been performed by varying α between 1 and 100. Bayesian information criterion is used to fit the model.

Due to the impressive learning capability of ConvAE, further experiments were performed to combine them with the cluster information provided by DPGMM. MFCC feature vectors were clustered using DPGMM. In order to train ConvAE, input-output pairs were randomly selected from the same cluster. Assuming that DPGMM clusters the samples based on similarity in manner of articulation (Table 1), input-output pair represent samples belonging to same

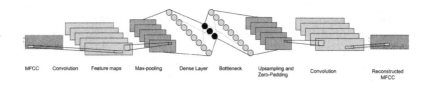

Fig. 1. Convolutional autoencoder architecture

articulatory gesture. This experiment will perform better if the representation being clustered preserves gestural information. We obtain cluster information by modeling MFCCs and ConvAE bottleneck activations using DPGMM. Results indicate that ConvAE bottleneck activations preserve gestural information and form clusters corresponding to similar articulatory gestures.

Table 1. Phoneme groups based on manner of articulation

Manner of articulation	Phonemes
Front vowels	/ih/, /iy/, /eh/
Mid vowels	/aa/, /er/, /ah/, /ao/
Back vowels	/uh/, /uw/, /ow/
Diphthongs	/ay/, /aw/, /ey/
Liquid semivowels	/w/, /l/, /el/
Glide semivowels	/r/, /y/
Nasals	/m/, /n/, /ng/, /nx/, /en/
Voiced stops	/b/, /d/, /dh/, /dx/, /g/
Voiced fricatives	/v/, /th/, /z/
Unvoiced fricatives	/f/, /s/, /sh/
Whisper	/hh/
Affricatives	/jh/, /ch/

2.3   Manifold Learning

Much of the acoustic variations in similar speech sounds can be reduced by projecting the features into a low dimensional representation where tangent directions of manifold are well preserved. PCA is a linear manifold learning technique where principal components point to the intrinsic coordinates of the manifold. As stated earlier, linear manifold learning techniques are not very successful to obtain meaningful representation of real-world data. Some of the non-linear manifold techniques assume that points may lie on a curved surface in a vector space which follows a Riemann geometric structure, hence they can be associated together using geodesic distance instead of straight-line distance. This section will describe some of the manifold learning techniques we investigated to discover the underlying geometric structure inherent to speech sounds.

Locally Linear Embedding. LLE is unsupervised dimensionality reduction technique where input is mapped into a single global coordinate system of lower dimensionality such that the distances of points in local neighbourhood are preserved. Standard LLE [23] can be summarized in following three steps:

1. **Nearest Neighbourhood Search:** A similarity measure is selected, based on which pairwise similarity is computed and k-nearest neighbours are chosen.

2. **Weight Matrix Construction:** Each point is reconstructed as a weighted sum of its neighbours. The weights are computed by minimizing the reconstruction error $E(W)$ given by:

$$E(W) = \sum_i ||X_i - \sum_j W_{ij}.X_j||^2 \qquad (8)$$

3. **Partial Eigenvalue Decomposition:** Low dimensional embeddings ($y_i \in \mathbb{R}^d$) of a test sample Y_i can be obtained by minimizing the cost function of the following form as below, which is similar to Eq. 8.

$$\Omega(W) = \sum_i ||Y_i - \sum_j W_{ij}.Y_j||^2 \qquad (9)$$

Embedding of Y_i is encoded in eigenvectors corresponding to d largest eigenvalues of $N \times N$ sparse matrix.

Isomap. Isomap [24] (Isometric Mapping) aims to preserve geodesic distances between all pair of points in space. It can be assumed as an extension of classical Multidimensional Scaling (MDS). Isomap approximates the geodesic distance between two points by finding shortest paths in the neighbourhood graph using Dijkstra's or Floyd-Warshall algorithm. A matrix D_G is constructed which stores shortest path between each pair of nodes in graph G. The d-dimensional embedding is obtained by using eigenvalues of matrix D_G. The p-th component of d-dimensional embedding vector y_i is given by:

$$y_i[p] = \sqrt{\lambda_p} v_p^i \qquad (10)$$

where λ_p is the p-th eigenvalue of matrix D_G and v_p^i is the i-th component of the p-th eigenvector of D_G.

Laplacian Eigenmaps. Laplacian eigenmaps [3] ensures the preservation of locality of samples on the low dimensional manifold. It uses the correspondence between graph Laplacian, Laplace Beltrami operator on the manifold and principal of transference of heat given by the heat equation.

We have tried other manifold learning methods like t-SNE and multidimensional scaling (MDS) but the results were inferior to other methods. Hence these methods are ommitted from the results. Though we have shown t-SNE visualization in Fig. 3 for the sake of comparison.

3   Experimental Setup

3.1   Data

We used Buckeye Corpus provided as a part of ZRSC for extracting phoneme samples and creation of a datset for benchmarking various dimensionality reduction techniques. Buckeye corpus is conversational speech recordings of 40 speakers conversing with an interviewer. The speech recordings are available at a sampling frequency of 16 kHz. We extracted 500 phoneme samples of duration more than 100 ms for 42 phoneme classes. Each phoneme segment is windowed at 25 ms and overlap of 10 ms to create frames. MFCC features with Δ and $\Delta\Delta$ were extracted for frame using Voicebox *melcepst* routine.

We use about 30 min of data for training autoencoder and manifold models which contains balanced amount of speech sounds representing 42 different phonemes.

3.2   Training

We divided the data such that 400 samples will be used for training and validating the models, while 100 samples will be used as test set.

Neural network architectures were implemented using Keras [9]. Each autoencoder model was trained for 50 epochs with early stopping based on validation loss. The bottleneck layer which is essentially the middle layer activation is used as the learnt feature. It was varied from 1 to 13 for each model. Activation functions used for training were also played around with. Tanh and PReLU perform the best in each autoencoder model variation. Various learning algorithms were tested such as SGD, Adadelta, RMSProp and Adam. We noticed that Adam converges faster and yields best results in all the cases.

Standard autoencoders were trained with 60-n-60 architecture, where n is the number of bottleneck units varied between 1 to 13. In case of ConvAE, input is convolved with 16 kernels of size 1×3 initialized with random gaussian weights, resulting in 16 feature maps. These feature maps are max-pooled with a kernel of size 1×2. Resulting feature maps are further convolved with 32 kernels of size 1×3 followed by max-pooling. The resultant of these two convolution is flattened and fed to a densely connected layer to neurons, which lead to bottleneck layer. Decoding is performed using unpooling (duplication of neighbouring activations) and convolution. The reconstructed output is compared to the input and weights are adjusted accordingly using appropriate learning law (in our case, Adam) to minimize the reconstruction error.

We re-used the weights obtained after training the convAE models to initialize the weights of denoising convAE. We added artificial Gaussian noise of zero mean and 0.2 standard deviation to input samples. Denoising convAEs do not help enhance the performance hence are ommitted from the results.

Manifold learning algorithms were implemented in Python using scikit-learn [21] library. Initially each of the manifold models were trained with 400 samples/phoneme extracted from Buckeye corpus for 42 phonemes, keeping number of neighbours to reconstruct the samples as 14. Each model was trained to transform input to n-dimensional space, where n varies from 1 to 13. 100 samples/phoneme were kept out of training to ensure generality while testing. After the observations from vowel classification task using manifold transformation, Isomap was singled out as the most discriminative algorithm for transforming speech inputs.

Generally, training time for manifold learning algorithms varies based on the target dimensionality. Autoencoders exhibit similar training time regardless of bottleneck dimension. Table 2 presents training time for each algorithm to train a single model. Dataset remains common for each algorithm.

Table 2. Time taken for each algorithm to train a single model

Algorithm	Training time (min)
LLE (standard, LTSA, Modified, Hessian)	30–40
Isomap	11–40
Laplacian Eigenmaps	18–45
Standard AEs (50 epochs)	∼10
Convolutional AEs (50 epochs)	∼25

3.3 Evaluation

We validate our results using two evaluation criteria. Preliminary evaluation helps us deduce the unsupervised technique best suited to undo the effect of speaker-dependent and extra-linguistic information in speech unit. This will result in clustering of similar sounds together and promote separability of dissimilar classes. Support Vector Machines (SVMs) were employed to validate the mentioned hypothesis of separability. Radial basis function (RBF) was used as kernel for SVM. For the sake of comprehensible analysis classification tests were conducted for 5 phonemes, although models were trained for all the 42 phonemes. Comparisons were made between 5 phoneme classes, namely /aa/, /iy/, /uw/, /eh/, /ae/. All the obtained transformed features were subjected to this evaluation and their results are compared.

We further conduct ZRSC's ABX discriminability test over a set of selected representations obtained from above mentioned models which perform exceptionally well in vowel classification task. According to ABX task, given a sample X determine its category based on its distance from A or B, where A and B belong to different categories. This measure is calculated in two different settings. Within speaker ABX measures discrimination of sample X when A, B and X belong to the same speaker. Across speaker ABX measures discrimination of sample X when A and B belong to a different speaker than X. Distance measure is chosen as cosine distance which is used when representation are compared using DTW. The final error rate is computed as the mean of error over all (A, B, X) in the test set.

4   Results and Discussion

Results are presented with respect to experiments performed with each low-dimensional manifold learning technique. Models with various parameters have

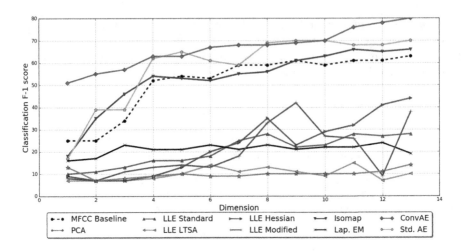

Fig. 2. Vowel Classification Accuracy obtained for each representation

been learnt for each one of them. We present the results of a selected set of models which show a contrast of performance between different methods. Figure 2 shows a comparison graph between all the techniques in vowel classification task. X-axis represents the dimensionality of manifold projection and Y-axis represents the F-1 score obtained for using a Kernel SVM classifier. MFCC features are represented by dotted black curve. We observe clearly that standard autoencoders and Isomap almost always perform better than MFCCs except in few high dimensional regions where they begin to converge. On the contrary convAEs consistently perform better than all other methods and continue to gain performance in higher dimensions. Other manifold learning techniques perform inferior to MFCCs in all the cases. But that may be only the case with phoneme classification. With further analysis, it can be verified whether these representations perform better for other properties like emotions classification or speaker verification, etc.

In order to visualize the separation of phonemes regions in 2D space, we plotted 3 vowels namely /aa/, /iy/, /uw/ using a scatter graph illustrated by Fig. 3. We can see that standard AEs and convAEs are superior in clustering the phone regions together as compared to other manifold learning methods. Figure 4 shows input and learned representation of the same phoneme by 2 different speakers. We notice a clear visual similarity between the representations, which suggests that ConvAEs were able to learn the invariance of similar sounding phonemes.

We also use minimal pair ABX discrimination task to show that using a handful amount of carefully selected data can be sufficient to introduce prior information inside the model which learns to discriminate between different speech sounds. Results show that autoencoder architectures and Isomap perform impressively well in such cases. Figure 5 presents a graph showing the

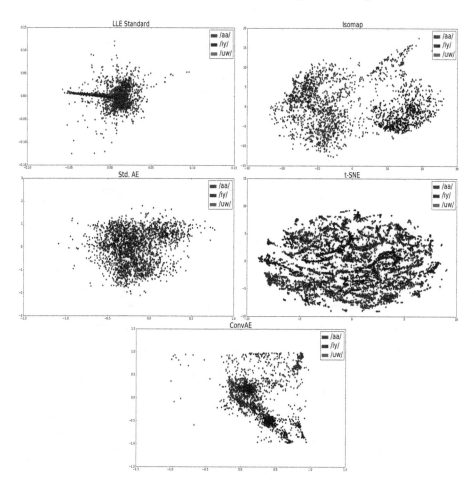

Fig. 3. Comparison of 3 vowels (/aa/ in blue, /iy/ in green, /uw/ in red) as represented by 2-dimension embeddings obtained from Std. LLE, Isomap, Std. AE, t-SNE and ConvAE top to bottom, left to right (Color figure online)

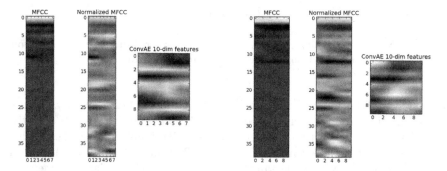

Fig. 4. Invariance of 10-dim representation obtained from ConvAE for vowel (/aa/) segments from 2 different speakers

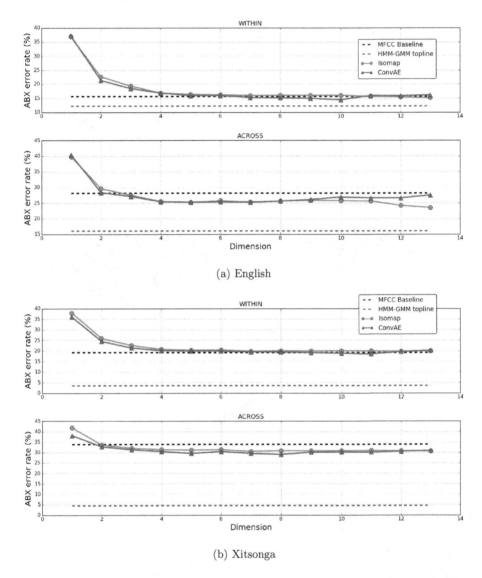

(a) English

(b) Xitsonga

Fig. 5. ABX discriminability error rates obtained over English and Xitsonga corpus using Isomap (yellow) and ConvAE (red) (Color figure online)

performance of both the algorithms as a function of dimensionality. Across-speaker discrimination is well induced even at 3-dimensions with 10–16% relative improvement, though within-speaker discrimination barely crosses the baseline with 3–4% relative improvement. This is possibly due to insufficient top-down constraints induced by training samples. To solve this issue, bigger segments of recurring units must be discovered in an unsupervised manner, which can be used as better top-down constraints (Fig. 6).

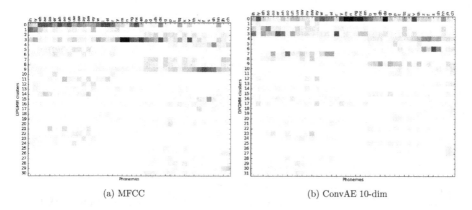

(a) MFCC   (b) ConvAE 10-dim

Fig. 6. Plots showing DPGMM clusters obtained using MFCC features and ConvAE 10-dim bottleneck features. X-axis represents the phonemes and Y-axis represents the DPGMM clusters. Phonemes are strategically placed closer based on their manner of articulation (refer Table 1). Images are greyscale where value of each column is sum-normalized. Black cells are closer to 1 and white ones are closer to 0. It's noteworthy from the ConvAE plot that clusters are better grouped in correspondence to phoenemes with similar manner of articulation as compared to MFCC. Note that there are different number of clusters for both representations since they are automatically chosen based on α parameter of DPGMM and inherent data distribution

5   Conclusion and Future Direction

We have presented the results of our experiments with an array of dimensionality reduction techniques in order to investigate the manifold hypothesis. Each of the representations were evaluated using phoneme classifcation and ABX discrimination task. In order to demonstrate the discriminability between different sounds and invariance of similar sounds manifested by these representations, all the models were trained by a common dataset created with a mix of balance set of phonemes. Among the geometric manifold learning techniques, Isometric Mapping outperforms MFCCs on both the tasks suggesting that similar set of phoneme sounds lie of a curved manifold inside higher-dimensional MFCC space where minimum geodesic distance is more intuitive than straight line euclidean distance.

Results indicate that autoencoders in general learn better representations than geometric manifold learning techniques discussed within the scope of this work. Performance of ConvAE is superior in both the tasks demonstrating that they are capable of learning better representations than other transformations and cluster them in highly separable regions even in 2-dimensions. ConvAEs can provide a highly robust representation of speech sounds by using very few parameters as compared to deep autoencoder architectures proposed in earlier works. Different representations were also clustered using DPGMM in order to learn more robust articulatory features. Clusters obtained using ConvAE representations display high correlation with articulatory classes. This cluster information was utilized to simulate correspondence term for ConvAE training which can be potentially used as fully unsupervised learning of articulatory phonological units. As a future direction of this work, we plan to explore recurrent autoencoder models in order to capture the temporal dependencies of articulatory gestures.

Acknowledgments. Authors would like to thank ITRA, Media Lab Asia to fund this research.

References

1. Anguera, X., Dupoux, E., Jansen, A., Versteegh, M., Schatz, T., Thiollière, R., Ludusan, B.: The zero resource speech challenge
2. Badino, L., Mereta, A., Rosasco, L.: Discovering discrete subword units with binarized autoencoders and hidden-Markov-model encoders. In: Sixteenth Annual Conference of the International Speech Communication Association (2015)
3. Belkin, M., Niyogi, P.: Laplacian eigenmaps for dimensionality reduction and data representation. Neural Comput. **15**(6), 1373–1396 (2003)
4. Blumstein, S.E., Stevens, K.N.: Acoustic invariance in speech production: evidence from measurements of the spectral characteristics of stop consonants. J. Acoust. Soc. Am. **66**(4), 1001–1017 (1979)
5. Browman, C.P., Goldstein, L.: Articulatory gestures as phonological units. Phonology **6**(02), 201–251 (1989)
6. Browman, C.P., Goldstein, L.: Articulatory phonology: an overview. Phonetica **49**(3–4), 155–180 (1992)
7. Browman, C.P., Goldstein, L.: Dynamics and articulatory phonology. In: Port, R.F., van Gelder, T. (eds.) Mind as Motion, pp. 175–193. MIT Press, Cambridge (1995)
8. Browman, C.P., Goldstein, L.M.: Towards an articulatory phonology. Phonology **3**(01), 219–252 (1986)
9. Chollet, F.: Keras (2015). https://github.com/fchollet/keras
10. Errity, A., McKenna, J.: An investigation of manifold learning for speech analysis. In: INTERSPEECH. Citeseer (2006)
11. Görür, D., Rasmussen, C.E.: Dirichlet process gaussian mixture models: choice of the base distribution. J. Comput. Sci. Technol. **25**(4), 653–664 (2010)
12. Greenberg, S., Kingsbury, B.E.: The modulation spectrogram: in pursuit of an invariant representation of speech. In: 1997 IEEE International Conference on Acoustics, Speech, and Signal Processing, ICASSP 1997, vol. 3, pp. 1647–1650. IEEE (1997)
13. He, K., Zhang, X., Ren, S., Sun, J.: Delving deep into rectifiers: Surpassing human-level performance on imagenet classification. In: Proceedings of the IEEE International Conference on Computer Vision, pp. 1026–1034 (2015)
14. Kamper, H., Elsner, M., Jansen, A., Goldwater, S.: Unsupervised neural network based feature extraction using weak top-down constraints. In: 2015 IEEE International Conference on Acoustics, Speech and Signal Processing (ICASSP), pp. 5818–5822. IEEE (2015)
15. Kingma, D., Ba, J.: Adam: a method for stochastic optimization (2014). arXiv preprint arXiv:1412.6980
16. Lee, C.-H., Siniscalchi, S.M.: An information-extraction approach to speech processing: analysis, detection, verification, and recognition. Proc. IEEE **101**(5), 1089–1115 (2013)

17. Leng, B., Guo, S., Zhang, X., Xiong, Z.: 3D object retrieval with stacked local convolutional autoencoder. Sig. Process. **112**, 119–128 (2015)
18. Makhzani, A., Frey, B.J.: Winner-take-all autoencoders. In: Advances in Neural Information Processing Systems, pp. 2773–2781 (2015)
19. Masci, J., Meier, U., Cireşan, D., Schmidhuber, J.: Stacked convolutional auto-encoders for hierarchical feature extraction. In: Honkela, T. (ed.) ICANN 2011, Part I. LNCS, vol. 6791, pp. 52–59. Springer, Heidelberg (2011)
20. Ostendorf, M.: Moving beyond the 'beads-on-a-string' model of speech. In: Proceedings of the IEEE ASRU Workshop, pp. 79–84. Citeseer (1999)
21. Pedregosa, F., Varoquaux, G., Gramfort, A., Michel, V., Thirion, B., Grisel, O., Blondel, M., Prettenhofer, P., Weiss, R., Dubourg, V., Vanderplas, J., Passos, A., Cournapeau, D., Brucher, M., Perrot, M., Duchesnay, E.: Scikit-learn: machine learning in Python. J. Mach. Learn. Res. **12**, 2825–2830 (2011)
22. Renshaw, D., Kamper, H., Jansen, A., Goldwater, S.: A comparison of neural network methods for unsupervised representation learning on the zero resource speech challenge. In: Proceedings of the Interspeech (2015)
23. Roweis, S.T., Saul, L.K.: Nonlinear dimensionality reduction by locally linear embedding. Science **290**(5500), 2323–2326 (2000)
24. Tenenbaum, J.B., Langford, J.C., De Silva, V.: A global geometric framework for nonlinear dimensionality reduction. Science **290**(5500), 2319–2323 (2000)
25. Tomar, V.S., Rose, R.C.: Application of a locality preserving discriminant analysis approach to ASR. In: 2012 11th International Conference on Information Science, Signal Processing and their Applications (ISSPA), pp. 103–107. IEEE (2012)
26. Tomar, V.S., Rose, R.C.: Efficient manifold learning for speech recognition using locality sensitive hashing. In: 2013 IEEE International Conference on Acoustics, Speech and Signal Processing (ICASSP), pp. 6995–6999. IEEE (2013)
27. Tomar, V.S., Rose, R.C.: Noise aware manifold learning for robust speech recognition. In: 2013 IEEE International Conference on Acoustics, Speech and Signal Processing (ICASSP), pp. 7087–7091. IEEE (2013)
28. You, M., Chen, C., Bu, J., Liu, J., Tao, J.: Emotional speech analysis on nonlinear manifold. In: 18th International Conference on Pattern Recognition, ICPR 2006, vol. 3, pp. 91–94. IEEE (2006)

Estimating the Severity of Parkinson's Disease Using Voiced Ratio and Nonlinear Parameters

Dávid Sztahó$^{(\boxtimes)}$ and Klára Vicsi

Laboratory of Speech Acoustics,
Department of Telecommunication and Media Informatics,
Budapest University of Technology and Economics,
Magyar Tudósok körútja 2, Budapest 1117-H, Hungary
{sztaho,vicsi}@tmit.bme.hu

Abstract. Parkinson's disease severity estimation analysis was carried out using speech database of Spanish speakers. Correlation measurements were performed between acoustic features and the UPDRS severity. The applied acoustic features were the followings: voicing ratio (VR), nonlinear recurrence: the normalized recurrence probability density entropy (H_{norm}) and fractal scaling: the scaling exponent (α). High diversity is found according to the type of speech sound production, and hence according to the text and gender. Based on the results of correlation calculations, prediction of the UPDRS values was performed using regression technique applying neural networks. The results showed that the applied features are capable of estimating the severity of the PD. By assigning the mean predicted UPDRS for each corresponding speaker using the best correlated linguistic contents, the result of the Interspeech 2015 Sub-challenge winner was exceeded. By training NN models separately for males and females the accuracy was further increased.

Keywords: Speech analysis · Parkinson's disease · Correlation analysis · Regression

1 Introduction

Parkinson's disease (PD) is one of the most common neurodegenerative disorders with an incidence rate of approximately 20/100 000 [16]. Main cause of Parkinson's disease is the damage or death of neurons that produce dopamine in the brain area called substantia nigra. Dopamine is a neurotransmitter that participates in the smooth, flicker-free, fluent regulation of skeletal muscles (muscles that execute voluntary movements). The main risk factors of PD are age, genetics and medicals, but the main most important factor is the age. This fact is rather important because in a continuously aging society the number of patients may also increase.

Main symptoms of PD include tremor, rigidity and loss of muscle control in general, as well as cognitive impairment. During diagnosis the therapist evaluates the patient's mental state, muscle strength, movement coordination, reflexes, and

© Springer International Publishing AG 2016
P. Král and C. Martín-Vide (Eds.): SLSP 2016, LNAI 9918, pp. 96–107, 2016.
DOI: 10.1007/978-3-319-45925-7_8

sensory skills. However, researches [12,18] show that speech may be a useful information for discriminating patients of PD from healthy controls. Clinical evidence suggests that most PD patients have some form of speech disorders [9]. In fact, speech can be an early sign of PD [8]. Informal description of symptoms that can characterize speech of PD patients are reduced loudness, increased vocal tremor, breathiness. The two vocal impairments linked to PD are dysphonia (inability to produce normal vocal sounds) and dysarthria (difficulty in pronouncing words) [2].

The reasons of diagnosing PD from speech at an early stage inspired researchers to develop decision support tools in order to identify PD. Various speech features were used that try to obtain relevant information about PD and help the differentiation between speech of PD patients and healthy control. In [20] a set of features is assigned into different classes of dysphonia measures.

Decision support tools discriminate PD patients from healthy controls by means of automatic classification methods. Many classifiers are available with different distinctive strength. Most commonly statistical classifiers are used [1,6,17,20], such as Bayesian classifier, support vector machines, (deep) neural networks, gaussian mixture models, random forests, k-nearest neighbors. In [20] binary classification was performed on sustained vowels only for two classes: PD and healthy control. In another study [10] about 85 % highest accuracy was achieved with speech intelligibility features from running speech using three classes: healthy, mild and severe.

In the present paper instead of categorizing patients into two groups (healthy, ill), the severity of the PD is to be estimated on a continuous scale.

Due to tremor and muscle movement disorders patients with PD show difficulty in producing fast voiced and unvoiced speech segment repetitions. They can't stop, or have difficulty in stopping voicing. This suggests that the main acoustic features that characterizes the syndrome are mainly related to parameters that measures voicing quality, for example in a close area of unvoiced plosives followed by or before vowels (VCV connections, where the consonant is an unvoiced sound).

Vocal production has nonlinear dynamics with certain randomness, and any changes in muscles and nerves will affect both the stochastic and deterministic components of the system. Rahn and his colleague have shown that aperiodic segments which are perceived as hoarse or breathy phonation, have an elevated incidence and are more prevalent in PD subjects [15]. Nonlinear recurrence and fractal scaling [13] are two properties that is successfully applied to voice disorder detection (differentiation between healthy and pathological speech).

On the base of the before mentioned two works we hypothesized that nonlinear recurrence and fractal scaling features and the severity of the Parkinson's disease (UPDRS score) may also correlate. Thus in our work we investigate how nonlinear recurrence and fractal scaling parameters, with adding the ratio of voiced and unvoiced speech segments, contribute to the severity prediction of PD.

In the Sect. 2 the applied database is described. In Sect. 3 the acoustic features are detailed. In the following Sections, first the correlation between the measurements of each feature and the UPDRS values are presented separately, then regression was performed to predict PD severity (UPDRS) using neural networks. In Sect. 6 the results are discussed and in the Session 7 conclusion and plans for the future are given.

2   Database

The database of the speech of patients with Parkinson's disease was created by Orozco-Arrovave et al. [14]. The database was used as the common database for Interspeech 2015 Special Session: PC Sub-challenge [19]. It contains speech in Spanish from 50 people (25 male, 25 female) suffering from PD. The age of the male utterances ranges from 33 to 77 (mean 62.2 ± 11.2), the age of the female utterances ranges from 44 to 75 (mean 60.1 ± 7.8). The data comprises a total of 42 speech tasks per speaker, including 24 isolated words, 10 sentences, one read text, one monologue, and the rapid repetition of syllables. The total duration of the recordings is 1.4 hours. The neurological state of the patients was evaluated by an expert neurologist according to the unified Parkinson's disease rating scale (motor subscale): UPDRS-III. The values of the neurological evaluations performed over the patients range from 5 to 92. The audio files are divided into training and development sets. An additional test partition with labels unknown to the participants comprises a further eleven subjects. In the present study the training and development sets of the database are used due to their additional meta-information (speaker, linguistic content) and test dataset was omitted (due to missing meta-information).

A more detailed description of the database can be found in [14].

3   Methods

Patients suffering from Parkinson's disease have disordered muscle movements, therefore supposedly having trouble starting and stopping vocal cord vibrations. This may be reflected at starts and stops of voiced speech segments. At these parts, such as VCV connections, where unvoiced consonants are present the unvoiced part becomes voiced due to the incorrect constant voicing. This can be measured with the ratio of the voiced and unvoiced parts of the speech. This ratio is highly linguistic and text dependent. A text in which the alternation of voiced and unvoiced sounds is frequent, this ratio differs more between the speech of PD patients and healthy control.

In order to calculate the voiced ratio (VR) the following equation was used:

$$VR = \frac{\sum_{i=1}^{n_v} d_v^i}{d_s} \tag{1}$$

where d_v^i is the duration of voiced segment i and d_s is the total duration of the speech in the sample. In order to calculation VR, first the speech samples were segmented into speech and non-speech parts using MAUS [3], then voiced parts were determined using Voicebox for Matlab [4]. Fundamental frequency was calculated with 10 ms time step and 40 ms window length.

Traditional voice quality measures, such as jitter and shimmer capture the quality of the voice production. However, during voice production there is a combination of deterministic and stochastic elements. The deterministic element is attributable to the nonlinear movement of the vocal fold and to the bulk of air in the vocal fold, whereas stochastic components are the high frequency aeroacoustics pressure fluctuations caused by the vortex shedding at the top of the vocal folds, whose frequency and intensity is modulated by the bulk air movement [13]. Time-delay embedding is a measurement for recurrence that captures this non-linearity. For a detailed description about the concept of recurrence, see [13]. Nonlinear recurrence (H_{norm}) was calculated using the Matlab tool of Max Little [13]. The recurrence time probability density was calculated by

$$P(T) = \frac{R(T)}{\sum_{i=1}^{T_{max}} R(i)},$$ (2)

where $R(\bullet)$ are the recurrence times, T_{max} is the maximum recurrence time found in the embedded time-state space. The normalized recurrence probability density entropy (RPDE) scale is computed by

$$H_{norm} = \frac{-\sum_{i=1}^{T_{max}} P(i) \ln P(i)}{\ln T_{max}}$$ (3)

Another aspect is the increased breath noise in disordered speech. A practical, robust approach to measure this phenomenon is the detrended fluctuation analysis (DFA) [13]. DFA scaling exponent (α) was calculated with [11]. First, the time series s_n is integrated:

$$y_n = \sum_{j=1}^{n} s_j,$$ (4)

for n = 1, 2 ... N, where N is the number of samples in the signal. Then, y_n is divided into windows of length L samples. A least-squares straight line local trend is calculated by analytically minimizing the squared error E^2 over the slope and intercept parameters a and b:

$$\underset{a,b}{\mathrm{argmin}}\, E^2 = \sum_{n=1}^{L} (y_n - a_n - b)^2$$ (5)

Next, the root-mean-square deviation from the trend, the fluctuation, is calculated over every window at every time scale:

$$F(L) = \left[\frac{1}{L} \sum_{n=1}^{L} (y_n - a_n - b)^2 \right]^{\frac{1}{2}}$$ (6)

This process is repeated over the whole signal at a range of different window sizes L, and a log-log graph of L against F(L) is constructed. A straight line on this graph indicates self-similarity expressed as $F(L) \propto L^{\alpha}$. The scaling exponent α is calculated as the slope of a straight line fit to the log-log graph of L against F(L) using least-squares as above.

Both features H_{norm} and α was calculated at voiced segments of speech samples. For the determination of the voiced parts Voicebox was applied.

4   Correlation Measurements

The main task in the detection of Parkinson's disease is the estimation of its severity. Based on medical employees' subjective opinions, disorders in speech may precede other muscle tremor symptoms, therefore they can be a source of severity estimation in an early stage of the disease. The correlations between the measured features and the UPDRS scores were calculated using SPSS [5].

The Spearman correlations are summarized in Table 1. Relatively low correlation is measured. Only DFA feature (α) shows a slightly higher correlation. However, it is still considered as low.

Table 1. Spearman correlations between features and UPDRS scores on the total dataset.

Variable pairs	Spearman
UPDRS - VR	.215
UPDRS -α	.260
UPDRS -H_{norm}	−.061

Because Parkinson's disease has influence mainly on muscle movement, those speech parts that have the most variability in vocal cord loads may show higher correlation with the measured data. To see the effect of the different linguistic contents on the correlation between the acoustic features and the UPDRS values, each sample group with different linguistic content was examined separately. Correlations are shown in Table 2. Significant values are marked by * (p < 0.05) and ** (p <0.01).

Table 3 shows the Spearman correlation values in the case of UPDRS – VR and UPDRS – α pairs for words that has the top six correlation ($\rho > 0.380$). In the case of the VR feature the top six correlated samples all contain VCV parts with unvoiced consonant. This suggests that VR may be able to capture if unvoiced sounds become voiced due to PD. Table 3 shows the Spearman correlation values between UPDRS and α values also. α shows higher correlation not only in the above mentioned VCV segments, but where the whole sample is pronounced as voiced (not counted if unvoiced parts are at the beginning or at the end), thus giving information from voiced segments.

Table 2. Spearman correlations of samples with different textual content for undivided, male and female datasets. Linguistic groups are marked by the filename notation of the database. ('*': p<0.05; '**': p<0.01)

	UPDRS – VR			UPDRS – α			UPDRS - H_{norm}		
	Undivided	Male	Female	Undivided	Male	Female	Undivided	Male	Female
"apto"	0.236	0.275	0.034	0.426**	0.647**	0.299	−0.020	−0.072	0.171
"atleta"	0.146	0.299	−0.093	0.379*	0.589**	0.249	−0.380	−0.354	−0.223
"blusa"	0.273	0.443*	−0.309	0.331	0.418	0.281	0.093	0.092	0.248
"bodega"	0.023	0.078	−0.198	0.194	0.229	0.240	0.266	0.350	0.322
"braso"	0.397*	0.555**	0.173	0.060	0.270	−0.310	0.219	0.246	−0.062
"campana"	0.134	0.208	0.013	0.301*	0.358	0.301	−0.022	0.100	−0.160
"caucho"	0.448**	0.576**	0.190	0.250	0.533*	0.177	−0.130	0.066	−0.353
"clavo"	−0.137	−0.394	0.220	0.380	0.424	0.302	−0.112	−0.051	−0.124
"coco"	0.499**	0.610**	0.140	0.350*	0.243	0.652**	−0.146	−0.388	0.288
"crema"	−0.032	−0.044	−0.024	0.428**	0.419	0.453	−0.131	−0.256	0.070
"drama"	0.019	0.089	0.011	0.318*	0.476*	0.246	0.149	0.131	0.209
"flecha"	0.510**	0.739**	−0.206	0.490**	0.454*	0.305	−0.276	−0.435*	0.141
"gato"	0.382*	0.622**	0.035	0.116	0.282	−0.015	0.049	−0.156	0.363
"globo"	0.204	0.202	0.187	0.441**	0.442*	0.544*	−0.153	−0.181	−0.126
"grito"	0.284	0.451*	0.115	0.245	0.121	0.499*	0.096	0.252	−0.125
"ka"	0.221	0.284	0.213	0.278	0.433*	0.172	−0.105	−0.155	0.213
"llueve"	−0.020	−0.003	−0.182	0.086	0.083	0.146	−0.048	−0.110	0.257
monologue	0.304*	0.530**	−0.045	0.445**	0.389	0.566**	0.142	0.156	0.187
"name"	0.237	0.612**	−0.209	0.229	0.171	0.293	0.231	0.355	−0.012
"pa"	0.397*	0.543*	−0.029	0.323	0.248	0.326	−0.183	−0.195	−0.108
"pato"	0.298	0.515*	−0.096	0.299	0.440	0.433	0.035	0.214	−0.243
"plato"	0.276	0.437*	−0.036	0.197	0.439*	−0.203	−0.063	−0.017	−0.108
"presa"	0.164	0.299	0.001	0.119	0.290	0.016	0.037	−0.068	0.256
readtext	0.290*	0.651**	−0.264	0.330*	0.451*	0.241	−0.146	−0.092	−0.140
"reina"	0.242	0.309	0.102	0.322*	0.261	0.587**	−0.002	−0.028	0.033
"trato"	0.214	0.254	0.017	0.152	0.608**	−0.299	0.103	−0.010	0.327
"viaje"	0.196	0.269	0.122	0.355*	0.472*	0.212	−0.296	−0.320	−0.355
sentence1	0.287	0.539**	−0.238	0.127	0.227	0.213	−0.063	0.123	−0.213
sentence2	0.286	0.568**	−0.258	0.331*	0.220	0.438	−0.066	0.131	−0.235
sentence3	0.348*	0.622**	0.001	0.383*	0.376	0.453*	0.029	0.244	−0.200
sentence4	0.234	0.519*	−0.281	0.272	0.243	0.457*	−0.066	−0.004	−0.114
sentence5	0.386**	0.622**	−0.016	0.403**	0.412	0.360	−0.074	0.048	−0.148
sentence6	0.265	0.487*	−0.215	0.401**	0.425*	0.518*	−0.156	−0.089	−0.173
sentence7	0.226	0.452*	−0.081	0.425**	0.508*	0.550**	−0.158	−0.154	−0.107
sentence8	0.160	0.388	−0.320	0.231	0.357	0.284	−0.232	−0.201	−0.241
sentence9	0.359*	0.545**	−0.149	0.225	0.255	0.375	−0.092	−0.089	−0.064
sentence10	0.332*	0.589**	−0.124	0.101	0.008	0.291	−0.030	0.147	−0.154

Table 3. Spearman correlations between UPDRS – VR and UPDRS – α for words with the top six correlation values for different textual content in the cases of undivided, male and female datasets.

VR				α			
Text	Undivided	Male	Female	Text	Undivided	Male	Female
"flecha"	0.510**	0.739**	−0.206	"flecha"	0.490**	0.454*	0.305
"coco"	0.499**	0.610**	0.140	"globo"	0.441**	0.442*	0.544*
"caucho"	0.448**	0.576**	0.190	"crema"	0.428**	0.419	0.453
"braso"	0.397*	0.555**	0.173	"apto"	0.426**	0.647**	0.299
"papapa"	0.397*	0.543*	−0.029	"clavo"	0.380	0.424	0.302
"gato"	0.382*	0.622**	0.035	"atleta"	0.379*	0.589**	0.249

The gender of the speakers may be also an influential factor on the correlation. Therefore, the samples were divided into the two gender categories and the same correlation was calculated as in the previous case. Tables 2 and 3 also show the correlation values broken down according to the two different genders. Feature VR has higher correlation in the case of the male speakers, whereas it has no significant value among the female samples. In contrast, α has significant values in both cases.

H_{norm} seems to have less significant values than the other two features altogether. The VR and α values are depicted in function of UPDRS on Fig. 1 in the case of the word "flecha" uttered by male and female speakers, which shows high correlation in both features. The correlation is lower in the case of female speakers.

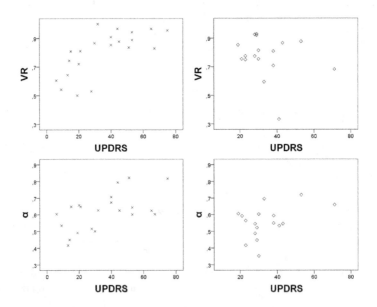

Fig. 1. VR (top) and α (bottom) feature values in function of UPDRS in the case of male (left) and female (right) samples of the word "flecha".

These phenomena suggest that the estimation of the severity of the Parkinson's disease should be performed separating speaker genders and taking only specific textual content into consideration.

5   Regression with Neural Network

Estimation of UPDRS values were carried out using feed-forward artificial neural networks (NN) (one hidden layer with 5 neurons). Based on the considerations of the previous Section, different test setups were applied: NN models were created using (1) all samples altogether, (2) samples with different linguistic content separately and (3) samples with separate gender and linguistic content. In each case 70 % of the speech samples were used for training NN models and the remaining 30 % was applied for testing. This database partitioning was identical to the one that was used at the Interspeech (IC) 2015 PC Sub-challenge (called training and development sets) [19]. The NN was implemented in SPSS. In cases (2) and (3) multiple models were trained and applied during prediction (the corresponding model for each speech sample). For each case the total prediction RMSE and correlation between the original and predicted UPDRS were calculated (Table 4). Beside Spearman correlation Pearson's r value is also depicted.

Table 4. Prediction errors between the original and the predicted UPDRS values. NN models were created using (1) all samples altogether, (2) samples with different textual content separately and (3) samples with separate gender and textual content. Baseline RMSE values are the standard deviations of the original UPDRS values. ('*': $p<0.05$; '**': $p<0.01$)

Predicted UPDRS calculation method		RMSE	Spearman	Pearson
Predicted UPDRS per utterance	Baseline	17.39	-	-
	(1)	16.33	0.303**	0.362**
	(2)	15.16	0.420**	0.498**
	(3)	14.14	0.540**	0.586**
Mean predicted UPDRS per speaker	Baseline	18.43	-	-
	(1)	15.82	0.574**	0.573**
	(2)	13.53	0.811**	0.760**
	(3)	12.25	0.846**	0.763**

Because each sample group with different linguistic content has samples from all the speakers, it is possible to create a single UPDRS value prediction for each speaker by averaging the NN outcomes. Table 5 includes the results with and without averaging the UPDRS values per speakers. Averaging resulted increased correlations, due to the negation of the error variance.

All three acoustic features were used to perform the regression. Although H_{norm} did not show significant correlation in Sect. 3, the regression tests showed that it improved the overall performance (with 0.02 in Spearman correlation).

Table 5. Prediction errors between the original and the mean predicted UPDRS per speaker using the best correlated linguistic content only. NN models were created using (1) all samples altogether, (2) samples with different linguistic content separately and (3) samples with separate gender and linguistic content. Baseline RMSE values (in brackets) are the standard deviation of the original UPDRS values. ('*': p<0.05, '**': p<0.01)

	Dataset	RMSE	Spearman	Pearson
(1)	Undivided	15.14 (18.43)	0.572*	0.626*
	Male	17.83 (22.76)	0.811*	0.710
	Female	12.32 (15.32)	0.252	0.552
(2)	undivided	13.79 (18.43)	0.717**	0.757**
	Male	15.80 (22.76)	0.811*	0.858*
	Female	11.75 (15.32)	0.571	0.656
(3)	Total	11.96 (18.43)	0.799**	0.831**
	Male	13.31 (22.76)	0.811*	0.844**
	Female	10.65 (15.32)	0.786*	0.891**
Interspeech 2015 baseline			0.492	-
Grósz et. al. [7]			0.691	-
Williamson et. al. [21]			0.670	0.671

Table 2 showed that not all of the sample groups (with different linguistic content) has the same effect on the performance of the UPDRS prediction. This suggests that the final prediction can be increased taking into consideration only those samples that has linguistic content with significant correlations (in the case of undivided dataset). The same test sessions were performed as in the previous case. The results are depicted in Table 5 showing mean predicted UPDRS values per speaker. The baseline value for the Sub-challenge (best performance on the developement set) is also depicted along with the results of the top two performing participants [7,21]. Figure 2 shows the best predicted and original UPDRS values with averaging per speakers in the best performing case.

6   Discussion

Based on the correlation analysis in Sect. 3 it is clear that the measured values are applicable to estimate the severity of the Parkinson's disease. However, not all of the speech samples have the same contribution, as well as not all the features. High diversity is found according to the type of speech sound production, linguistic content and text. For example, how many such VCV segments are present, where the consonant is unvoiced.

The measured correlation values of the male and female speakers show differences. Males tend to have more correlation between the computed features and the UPDRS values than females in the examined speech material. Whether is it due to a general tendency in the disease, or some other phenomena, it is not clear.

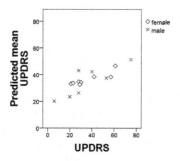

Fig. 2. Scatter plot of the original UPDRS and the predicted mean UPDRS values per speaker with NN models trained for male and female speakers separately using the best correlated linguistic content only.

In case of regression, the estimation of the UPDRS values showed that the applied features are capable of estimating the severity of the disease. The RMSE values were significantly lower than the standard deviation of the original UPDRS values (that is predicting the mean of UPDRS values for all utterances).

Because the database contained the speaker id values and each different linguistic content was uttered by all of the speaker, it was possible to create one predicted UPDRS for each speaker by averaging the results corresponding to the same speaker. This resulted much better UPDRS prediction by negating the variance of the prediction error. In a real world diagnostic application this can also be performed by recording utterances with varying linguistic content from the same speaker. Sorting out only the best correlated linguistic contents an even more accurate predictor could be trained. Based on the correlations of the individual features, only those were selected that had the highest values. With these units, performing the same prediction tests as in the previous case, increased correlations and lower RMSE values were achieved. The baseline result of the Interspeech 2015 Sub-challenge was surpassed. Although the predicted values have smaller range than the original ones (Fig. 2), the order (rank) of the UPDRS values remained mostly the same.

The female/male differences were observed here as well. This implied that training NN models with both gender types at the same time performs worse that separating them. The results in Tables 4 and 5 confirm this hypothesis. It is interesting to see that the performance increased more in the case of female samples compared to the male speakers. The reason of this is not clear.

7   Conclusion

In this paper a Parkinson's disease severity estimation analysis was carried out using speech database of Spanish speakers. Correlation measurements were performed between acoustic features and the UPDRS severity values to examine the goodness of the selected parameters. The applied acoustic features were the following: the voicing ratio (VR); nonlinear recurrence: the normalized recurrence probability density entropy (H_{norm}) and fractal scaling: the scaling exponent (α) were examined. High diversity is found according to the type of speech sound production, and hence according to the text and gender. Increased correlation was by separating male and female speech speakers. Males tend to have higher correlation between the computed features and the UPDRS values than females. The reason of this phenomenon is hereinafter should be thoroughly investigated.

Based on the results of correlation calculations, prediction of the UPDRS values was performed using regression technique applying neural networks. The results showed that the applied features are capable of estimating the severity of the PD. By assigning the mean predicted UPDRS for each corresponding speaker using the best correlated linguistic contents, the result of the Interspeech 2015 Sub-challenge winner was exceeded. It means, that by composing well selected, different linguistic content, more accurate predictor could be trained. Moreover, by training NN models separately for males and females the accuracy could be increased further in case of the examined Spanish database.

In the future our method must be tested by other databases beside this Spanish one. A Hungarian database of speech samples from patients suffering from Parkinson's disease are under recording, and will be complete at the end of 2016. The acoustic measurements on those samples may confirm the usability of the features described in this paper. Moreover, it will be appropriate to examine the extent of the contribution of several parameters, for accurate predictions. In case of success, a computer aided diagnostic system can be developed in helping medical staff or home self-diagnosis.

References

1. Alemami, Y., Almazaydeh, L.: Detection of Parkinson disease through voice signal features. J. Am. Sci. **10**(10), 44–47 (2014)
2. Baken, R.J., Orlikoff, R.F.: Clinical Measurement of Speech and Voice. Cengage Learning, Boston (2000)
3. Beringer, N., Schiel, F.: The quality of multilingual automatic segmentation using german maus. In: Proceedings of the International Conference on Spoken Language Processing (2000)
4. Mike Brookes. Voicebox (2016). http://www.ee.ic.ac.uk/hp/staff/dmb/voicebox/voicebox.html/
5. IBM Corp. Ibm spss statistics for windows, version 22.0 (2013)
6. Frid, A., Hazan, H., Hilu, D., Manevitz, L., Ramig, L.O., Sapir, S.: Computational diagnosis of Parkinson's disease directly from natural speech using machine learning techniques. In: IEEE International Conference on Software Science, Technology and Engineering (SWSTE), pp. 50–53. IEEE (2014)
7. Grósz, T., Busa-Fekete, R., Gosztolya, G., Tóth, L.: Assessing the degree of nativeness and Parkinson's condition using gaussian processes and deep rectifier neural networks. In: Proceedings of Interspeech, pp. 1339–1343 (2015)
8. Harel, B., Cannizzaro, M., Snyder, P.J.: Variability in fundamental frequency during speech in prodromal, incipient Parkinson's disease: a longitudinal case study. Brain Cogn. **56**(1), 24–29 (2004)

9. Ho, A.K., Iansek, R., Marigliani, C., Bradshaw, J.L., Gates, S.: Speech impairment in a large sample of patients with Parkinson's disease. Behav. Neurol. **11**(3), 131–137 (1999)

10. Khan, T., Westin, J., Dougherty, M.: Classification of speech intelligibility in Parkinson's disease. Biocybernetics Biomed. Eng. **34**(1), 35–45 (2014)

11. Little, M., McSharry, P., Moroz, I., Roberts, S.: Nonlinear, biophysically-informed speech pathology detection. In: IEEE International Conference on Acoustics Speech and Signal Processing Proceedings, vol. 2, p. II. IEEE (2006)

12. Little, M.A., McSharry, P.E., Hunter, E.J., Spielman, J., Ramig, L.O., et al.: Suitability of dysphonia measurements for telemonitoring of Parkinson's disease. IEEE Trans. Biomed. Eng. **56**(4), 1015–1022 (2009)

13. Little, M.A., McSharry, P.E., Roberts, S.J., Costello, D.A.E., Moroz, I.M.: Exploiting nonlinear recurrence and fractal scaling properties for voice disorder detection. BioMed. Eng. OnLine **6**(1), 1 (2007)

14. Orozco-Arroyave, J.R., Arias-Londoño, J.D., Bonilla, J.F.V., Gonzalez-Rátiva, M.C., Nöth, E.: New Spanish speech corpus database for the analysis of people suffering from Parkinson's disease. In: LREC, pp. 342–347 (2014)

15. Rahn, D.A., Chou, M., Jiang, J.J., Zhang, Y.: Phonatory impairment in Parkinson's disease: evidence from nonlinear dynamic analysis and perturbation analysis. J. Voice **21**(1), 64–71 (2007)

16. Rajput, M., Rajput, A., Rajput, A.H.: Epidemiology. In: Pahwa, R., Lyons, K.E. (eds.) Handbook of Parkinson's Disease, 4th edn. CRC Press, New York (2007)

17. Sakar, B.E., Isenkul, M.E., Sakar, C.O., Sertbas, A., Gurgen, F., Delil, S., Apaydin, H., Kursun, O.: Collection, analysis of a Parkinson speech dataset with multiple types of sound recordings. IEEE J. Biomed. Health. Inform. **17**(4), 828–834 (2013)

18. Sapir, S., Ramig, L.O., Spielman, J.L., Fox, C.: Formant centralization ratio: a proposal for a new acoustic measure of dysarthric speech. J. Speech Lang. Hear. Res. **53**(1), 114–125 (2010)

19. Schuller, B., Steidl, S., Batliner, A., Hantke, S., Hönig, F., Orozco-Arroyave, J.R., Nöth, E., Zhang, Y., Weninger, F.: The interspeech: nativeness, Parkinson's & eating condition. In: Proceedings of INTERSPEECH (2015)

20. Tsanas, A., Little, M.A., McSharry, P.E., Spielman, J., Ramig, L.O.: Novel speech signal processing algorithms for high-accuracy classification of Parkinson's disease. IEEE Trans. Biomed. Eng. **59**(5), 1264–1271 (2012)

21. Williamson, J.R., Quatieri, T.F., Helfer, B.S., Perricone, J., Ghosh, S.S., Ciccarelli, G., Mehta, D.D.: Segment-dependent dynamics in predicting Parkinson's disease. In: Sixteenth Annual Conference of the International Speech Communication Association (2015)

Optimal Feature Set and Minimal Training Size for Pronunciation Adaptation in TTS

Marie Tahon$^{(\boxtimes)}$, Raheel Qader, Gwénolé Lecorvé, and Damien Lolive

IRISA/University of Rennes 1, 6 Rue de Kérampont, 22300 Lannion, France
{marie.tahon,raheel.qader,gwenole.lecorve,damien.lolive}@irisa.fr
https://www-expression.irisa.fr/

Abstract. Text-to-Speech (TTS) systems rely on a grapheme-to-phoneme converter which is built to produce canonical, or statically stylized, pronunciations. Hence, the TTS quality drops when phoneme sequences generated by this converter are inconsistent with those labeled in the speech corpus on which the TTS system is built, or when a given expressivity is desired. To solve this problem, the present work aims at automatically adapting generated pronunciations to a given style by training a phoneme-to-phoneme conditional random field (CRF). Precisely, our work investigates (i) the choice of optimal features among acoustic, articulatory, phonological and linguistic ones, and (ii) the selection of a minimal data size to train the CRF. As a case study, adaptation to a TTS-dedicated speech corpus is performed. Cross-validation experiments show that small training corpora can be used without much degrading performance. Apart from improving TTS quality, these results bring interesting perspectives for more complex adaptation scenarios towards expressive speech synthesis.

Keywords: Speech synthesis · Pronunciation adaptation · Feature selection · Training data size

1 Introduction

Text-to-speech (TTS) systems mainly rely on two steps. First, the input text is converted to a canonical phoneme sequence using an automatic phonetizer. Then, the waveform is generated from this phoneme sequence by querying a dedicated database of speech segments or using generative models trained on this database. In such a framework, and as used in current TTS systems, phonemes generated by the phonetizer need to be consistent with those labeled in the speech corpus in order to produce high quality synthetic speech samples. Given an existing TTS application, this strong requirement makes it very difficult to change the phonetizer to another, move to a new speech database, or to consider expressive pronunciation variants, unless redesigning all the components from scratch. As a consequence, TTS applications are nowadays still grounded on a very limited variety of voices, yielding to culturally centered and neutrally accented systems [1]. One of the current challenges in TTS is thus to adapt

© Springer International Publishing AG 2016
P. Král and C. Martín-Vide (Eds.): SLSP 2016, LNAI 9918, pp. 108–119, 2016.
DOI: 10.1007/978-3-319-45925-7_9

standard pronunciations to new conditions, especially expressivity, speaking style or speaker characteristics [2].

To overcome this problem, our paper focuses on a new pronunciation adaptation method which adapts canonical phonemes generated by the phonetizer to a specific pronunciation style. Precisely, this method seeks here to adapt canonical phonemes to pronunciations uttered in the speech corpus on which the TTS system is built. Beyond the importance of this particular problem, our work is more generally regarded as a case study towards more complex adaptation scenarios, e.g., emotion or accent-specific adaptations. Using machine learning, the studied pronunciation adaptation method consists in training adaptation models on a target *pronunciation corpus*. In the perspective to deploy this method to various use cases, investigations are conducted in this paper on (i) the choice of optimal features and (ii) the minimal size of the pronunciation corpus to train reasonable adaptation models. Looking for an optimal feature set to model pronunciations is required to improve the adaptation accuracy without overfitting target pronunciations. It implies the addition, selection and combination of relevant linguistic, phonological, prosodic and articulatory features. Then, finding the minimal quantity of material needed for training reliable models is of first importance because the cost of casting, segmenting and annotating speech databases is still very high. To provide robust conclusions, this question is studied on a variety of feature configurations.

In recent literature, models of pronunciation have been proposed for both automatic speech recognition (ASR) and TTS. Many statistical approaches have already been used for pronunciation modeling. Among them, neural networks [3–5] conditional random fields (CRFs) [3,6,7] and bayesian networks [8] are the most frequent. In the present work, pronunciation is modeled with CRFs. Only few studies report experiments on the quantity of speech training material [9,10]. However, because the cost of data is still important it is necessary to evaluate data requirements in terms of size and content. Of course, statistics require large quantities of data, but in many fields of research –especially in affective computing– only small sized corpora are available, thus causing the problem of overfitting. In such cases, a compromise between both the quantity of training material and the size of the feature set needs to be reached. Whereas the search for data requirements has rarely been investigated, the search for an optimal feature set has been extensively studied, for example in the field of affective computing [11]. According to [12], with a small quantity of training material, reduced feature sets usually lead to models which better generalize than large feature sets. According to [13], "any subjective choice of which dimensions to keep and what heuristic reasoning to apply inevitably involves some assumptions about how the systems and workloads behave". As a consequence, a widely used method (also called brute-force method) begins with a large number of features, then performs dimension reduction. Another commonly used approach, set up in the present work, is to introduce some human knowledge to select features *a priori*.

The following work improves the method proposed by [14] and adapts it to a French speech corpus. CRFs will be trained with different feature sets and different quantities of training data. These experiments are able to estimate which differences between phonemes generated by a phonetizer, and phonemes from the speech corpus, can be fixed up with a small speech corpus. Apart from improving TTS quality, the presented pronunciation adaptation method brings interesting perspectives in terms of expressive speech synthesis.

In the remainder, the speech corpus, its derived features and the experimental set-up are introduced in Sect. 2. Features and phoneme window selection experiments are presented in Sect. 3. Section 4 presents the training data reduction protocol and its results. A pronunciation example is discussed in Sect. 5. Conclusion and perspectives are drawn in the last section.

2   Material and Method

This section is devoted to the presentation of the speech corpus used in the experiments, the description of the feature set and the presentation of the experimental set-up.

2.1   Speech Corpus

Experiments were carried out on a French speech corpus dedicated to interactive vocal system TTS. As such, this corpus covers all diphonemes present in French and comprises most used words in the telecommunication field. It features a neutral female voice sampled at 16 kHz (lossless encoding, one channel).

The corpus is composed of 7, 208 utterances, containing 196, 190 phonemes and 16, 750 non-speech sounds, totaling 5h49 of speech. Pronunciations and non-speech sounds have been strongly controled during the recording process. Other information has been automatically added and manually corrected. The corpus and its annotations are managed using the Roots toolkit [15].

2.2   Features

The goal of the present work is to reduce the differences between phonemes generated by the phonetizer during synthesis, referred to as *canonical phonemes*, and phonemes as labeled in the speech corpus, referred to as *realized phonemes*. To do so, the proposed method consists in training a CRF model which predicts corpus-specific phonemes from canonical ones. To enrich the model, and hopefully improve the prediction accuracy, other state-of-the-art features are added. Precisely, four groups of features have been investigated: linguistic, phonological, articulatory and prosodic features, thereby leading to 52 feature set adapted from [14]. Most features have been normalized to corpus or utterance and discretized.

Canonical phonemes are generated with Liaphon [16], one of the most widely used utterance phonetization system for French. Word frequencies in French are extracted from Google n-grams [17]. Articulatory features are standard International Phonetic Alphabet (IPA) traits. In an ideal system, prosody should also be predicted from text. However, because this task is still a research issue, prosodic features have been extracted in an oracle way, i.e., directly from the recorded utterances of the speech corpus. Such a protocol allows to know to what extent prosody affects pronunciation models. Prosodic features are based on energy, fundamental frequency (F_0) and duration. F_0 shape is based on a glissando value perceptually defined [18].

2.3   Experimental Set-Up

The phonemic sequences are modeled with CRFs, trained with the Wapiti toolkit [19]. Realized phoneme sequences and statistically adapted phoneme sequences are compared under the usual Phoneme Error Rate (PER). The speech corpus has been randomly split in two: training and development set (70 %), and a validation set (30 %). The training set has been divided in 7 folds. Models are trained on 6 folds, developed on 1 fold and tested on the remaining validation set. This protocol ensures that data used for training and testing do not overlap. The feature set at least consists in the canonical phoneme sequence generated with the phonetizer.

3   Optimal Feature Set

Finding an optimal feature set is a very important task in machine learning. It helps identify the feature subset which best predicts pronunciation, usually avoids overfitting the training data, and thus leads to models that generalize more to unseen data. Lastly, it reduces the time and memory required during the training process. In our method, features are selected for each group of features separately, using a forward selection process. Then groups of selected features are combined together with phoneme window to find the optimal configuration.

3.1   Feature Selection Within Groups of Features

Protocol. A forward feature selection protocol has been adapted to French from previous work on spontaneous English pronunciation [14]. A cross-validation selection process was performed on the initial training set (six folds for training, one for testing) without any phoneme window. For each group of features, the selection starts with canonical phonemes only and other features are added one at a time until the optimal subset is reached. In order to find the global subset from the seven subsets obtained for each fold, a voting process has been set up.

Selected Features. In the end, 15 linguistic, prosodic and phonological features were selected. Selected features are reported in the Table 1. First, it appears that two linguistic features were selected for all folds: the word itself and its stem.

Table 1. Selected features used for pronunciation modeling names and LPrPh feature set.

Group of feature	# feat	Selected features
Linguistic (L)	2	Word ♦ Stem
Phonological (Ph)	7	Canonical syllables ♦ Syllable in word position ♦ Phoneme reverse position in syllable (numerical) ♦ Phoneme position and reverse position (numerical) ♦ Word length in phoneme (numerical) ♦ Pause per Syllable (low, normal, high)
Articulatory (A)	0	-
Prosodic (Pr)	6	Syllable Energy (low, normal, high) ♦ Syllable and phoneme tone (from 1 to 5) ♦ F_0 phoneme contour (decreasing, flat, increasing) ♦ Speech rate (low, normal, high) ♦ Distance to previous pause (from 1 to 3)

Since these features are highly correlated, one would have expected only one feature to be selected. However, as stated in [20], "noise reduction and consequently better class separation may be obtained by adding variables that are presumably redundant". Word expectation features, such as word frequency in French, received only very few votes. Surprisingly, it appears that no articulatory features have reached the minimal number of votes. Since previous studies have shown the interest of such features for pronunciation variation modeling [8], they were expected to have better votes. Then, seven phonological features were included in the optimal set, most of them being related to phoneme positions in the utterance. None of the syllable characteristics (such as syllable part, structure or type) have been selected. Finally, six among seven prosodic features have been selected. This result is in agreement with state-of-the-art and suggests that a prosodic model is able to model a speaker's pronunciation.

3.2 Feature Group Combinations

Different combinations of selected feature groups were evaluated in cross-validation conditions, on the validation set without phoneme window. Average PER obtained on the seven folds are reported in bold in the Table 2. The baseline is the PER obtained without any adaptation, between phoneme sequence generated by the phonetizer and realized phoneme sequence (ground truth). An improvement of 4.6 % point (pp) is reached while using a pronunciation model with canonical phonemes only, thus showing how pronunciation adaptation can reduce the inconsistency between the phonetized output and the speech corpus. Separately adding groups of selected features further improves the PER. The most spectacular reduction lies in the linguistic group: with only two apparently redundant features (word and its stem), a drop of 6.8 pp is obtained from the baseline. Overall results show an improvement in PER when combining selected feature groups. The combination of prosodic and linguistic groups lead to a

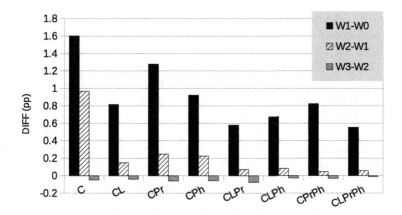

Fig. 1. Effect of the phoneme window size on the average PER obtained on 7 folds. $\text{DIFF}_i = \text{PER}(W_i) - \text{PER}(W_{i-1}), i \in \{1, 2, 3\}$.

significant drop in PER of 7.7 pp with a minimum number of features. The combination of the three feature groups brings the best PER, with an improvement of 7.9 pp from the baseline. In the end, only almost a third of the initial feature set remains.

3.3 Effect of Phoneme Window

In the search for an optimal feature set, a phoneme window is of real importance for pronunciation modeling since, linguistic, phonological and prosodic features of the current phoneme depend on the previous and next ones.

Protocol. Only symmetrical windows have been tested since asymmetric windows did not show any interesting improvements [14]. The application of a symmetrical phoneme window Wx ($2 \cdot x + 1$ phonemes) is performed on the current canonical phoneme, and also its associated features therefore multiplying the number of features in the CRF model by x. Four phoneme window sizes are tested under cross-validation conditions using the same protocol as in the previous section. The results in terms of averaged PER on the seven folds and for different feature combinations are reported in the Table 2. The relative gain obtained while increasing the window size is represented on Fig. 1 for different feature combinations.

Results. Figure 1 shows that adding features coming from one (black) or two (hatched) surrounding neighbours have a positive effect on the global PER. A seven phoneme window (W3) degrades the results, probably because as the number of feature increases, the model overfits the data. The effect of the phoneme window size differs according to the combination of feature used for training models. For instance, a phoneme window has a higher effect when models are trained with prosodic features than linguistic or phonological features. Indeed, prosodic features of the current phoneme highly depends on what precedes and what succeeds. Finally, according to Table 2, the combination of a window W2 and the 15 selected features brings the best results. In the next section, four configurations are tested: two phoneme window W0 and W2, and two feature sets: canonical phonemes only (C) and the 15 selected features with canonical phonemes (CLPrPh).

Table 2. PER values averaged on 7 folds in cross-validation conditions with adaptation. Different phoneme window sizes and different feature combinations. Baseline is 11.2%.

Window	C	CL	CPr	CPh	CLPr	CLPh	CPrPh	CLPrPh
W0	**6.6**	**4.4**	**4.8**	**4.5**	**3.5**	**4.0**	**3.7**	**3.3**
W1	5.0	3.6	3.6	3.6	2.9	3.3	2.8	2.8
W2	4.1	3.4	3.3	3.3	2.9	3.2	2.8	2.7
W3	4.1	3.5	3.4	3.4	3.0	3.2	2.8	2.7

Perceptive tests were realized on synthesized speech samples with different feature combinations [21]. As in [22], the results are strongly linked with PER and confirmed the relevancy of both the pronunciation adaptation model and the selected features. Some samples are available on the team website[1].

4   Minimal Training Data Size

Because the cost of pronunciation corpora is very high, it is worth trying to find the minimal quantity of training material required for pronunciation adaptation. The obtained results would tell us the expected accuracy for a given duration of training material. For a given quantity of training material, models are evaluated in terms of PER in cross-validation conditions.

4.1   Protocol

The training set (70% of the initial speech corpus) has been divided in $N_f = 7$ folds: 6 for training and the remaining for development purpose. The different size of training material is obtained while splitting the initial training set in 2×7, then 4×7, 8×7, etc. At each step, 6 folds are used for training models, one of the remaining folds is kept for development. In order to limit the experimental time, we have limited N_f to 100 for training durations less than 300 min While the quantity of training material decreases, N_f increases thus making results more reliable: from 243.3 min of training data ($N_f = 7$, 4321 utterances each) to 40 s of training data ($N_f = 100$, 12 utterances each). The validation set consists in 120.2 min of data and 2161 utterances. Two feature sets (C and CLPrPh) and

---

[1] http://www-expression.irisa.fr/demos/: Corpus-specific adaptation.

two phoneme windows (W0 and W2) are tested. This choice allows to study the effects of the number of features on accuracy and to estimate the danger of using too much features while training on too few data.

4.2 Results

As expected, the averaged results on Fig. 2 (top) show that the smaller the duration of training data, the higher the phoneme error rate. What is surprising is that models trained with very small data improve the PER of 4.0 pp. from baseline (best configuration W0-CLPrPh). Therefore, small training sets allow fixing many phoneme errors: recurrent pronunciation, alphabet mapping, French schwa and liaison (see Sect. 5). Of course, with very small training set, standard deviation computed on all the folds increase. Therefore, it appears that some sets allow to reach a good PER, whereas some others do not. Thus, the minimal quantity of training material does not lie in its duration only, but also in the content of this set.

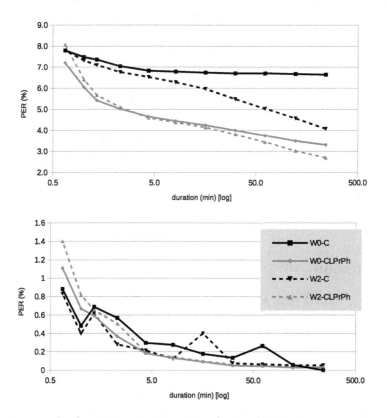

Fig. 2. Average (top) and standard deviation (bottom) PER between canonical and adapted phonemes obtained on the validation set. Average is computed on all available folds for a given training duration (log-scale minutes). Baseline is 11.2 %.

PER as a function of the duration of training data follows two different trends whether duration is over 4.4 min or not. Interestingly, for training duration over this threshold, the logarithmic curve of PER is almost linear with respect to duration (correlation coefficient over 0.96, see Table 3). This result is in agreement with the results obtained in ASR experiments [10]. For small durations (less than 4.4 min), phoneme window has almost no effect on PER whereas the number of feature does. When training models with very little data, there is a weak effect of window and feature sets (PER range is 0.9 pp.). The results show that CRF models trained with small datasets are still better than the baseline. All the more, multiplying the duration by 6.6 leads to an improvement of 2.6 pp. (best W0-CLPrPh configuration). For larger durations of training data (more than 4.4 min), feature set has less effect compared to phoneme window. Interestingly, increasing the amount of training data does not improve significantly the accuracy of models trained with W0-C configuration. In this case, multiplying the duration by 10 leads to an improvement of only -0.5 pp. (best W2-CLPrPh configuration).

The obtained results show that there is a threshold for duration of training data at almost 5 min. Over this threshold, the addition of new data has a high cost but a weak improvement in accuracy. Since the PER is log-linear with respect to the duration, an ideal PER = 0 would be reached for $3 \cdot 10^8$ hours of training data with W2-CLPrPh configuration.

Table 3. Linear regression results of PER w.r.t. training data duration (logarithmic scale).

Training duration	Lin. Reg	W0-C	W0-CLPrPh	W2-C	W2-CLPrPh
>0.7 min	Slope	−0.17	−0.54	−0.58	−0.73
	Corr. coef	0.74	0.85	0.99	0.86
>4.0 min	Slope	−0.04	−0.34	−0.62	−0.48
	Corr. coef	0.96	1.00	0.99	0.99

5   Discussion

CRF models trained with small datasets bring better results than the baseline in terms of PER, hence underlying the power of such pronunciation adaptation models. Models trained with very few utterances are able to fix some regular errors between canonical and realized phonemes: recurrent pronunciation, French schwa and liaisons. Generally, canonical phonemes and realized phonemes are not encoded using the same alphabet, therefore introducing phoneme differences which are not typical errors. Interestingly, CRF are able to solve the alphabet issues.

Table 4 shows an example of adaptation results on the pronunciation of an utterance in the validation set. This example illustrates typical errors. First, a one-phoneme window (W0) is not able to model French liaisons (in the example: /s ɔ̃ t ɛ/), whatever the duration of the training set. A larger phoneme window

Table 4. Example of pronunciation adaptations with different windows, features and training size. The input text is *Dans la montagne, les couleurs sont exceptionnelles.* "In the mountains, colors are remarkable"

Win.	Features	dur(min)	Phoneme sequence
Realized			d ã l a m ɔ̃ t a ɲ j - l e k u l œ ʁ s ɔ̃ t ɛ k s ɛ p s j o n ɛ l -
Canonical			d ã l a m ɔ̃ t a ɲ - ə l e k u l œ ʁ s ɔ̃ - ɛ k s ɛ p s j ɔ n ɛ l ə
W2	CLPrPh	243.3	d ã l a m ɔ̃ t a n j - l e k u l œ ʁ s ɔ̃ z e k s e p s j ɔ n ɛ l -
W2	C	243.3	d ã l a m ɔ̃ t a n j - l e k u l œ ʁ s ɔ̃ t e k s ɛ p s j o n ɛ l -
W0	C	243.3	d ã l a m ɔ̃ t a n j - l e k u l œ ʁ s ɔ̃ - ɛ k s ɛ p s j ɔ n ɛ l -
W2	CLPrPh	4.4	d ã l a m ɔ̃ t a n j ə l e k u l œ ʁ s ɔ̃ t ɛ k s ɛ p s j o n ɛ l -
W2	C	4.4	d ã l a m ɔ̃ t a n j - l e k u l œ ʁ s ɔ̃ t ɛ k s ɛ p s j o n ɛ l -
W0	C	4.4	d ã l a m ɔ̃ t a n j - l e k u l œ ʁ s ɔ̃ - ɛ k s ɛ p s j o n ɛ l -
W2	CLPrPh	0.7	d ã l a m ɔ̃ t a g - e l e k u l œ ʁ s ɔ̃ - ɛ k s ɛ p s j o n ɛ l -
W2	C	0.7	d ã l a m ɔ̃ t a ʁ - - l e k u l œ ʁ s ɔ̃ t ɛ k s ɛ p s j o n ɛ l -
W0	C	0.7	d ã l a m ɔ̃ t a ʁ - - l e k u l œ ʁ s ɔ̃ - ɛ k s ɛ p s j ɔ n ɛ l -

combined with C or CLPrPh is able to model the liaison, but the result is not always correct (/z/ instead of /t/ with 243.3 min of training data). CRF models trained with 40 s of data are not able to label correctly the canonical symbol /ɲ/: labels /n j/ are not found but /ʁ/ or /g/. The deletion of French schwa is realized in all configurations at the end of the utterance. Models trained with the full CLPrPh feature set and few data label the schwa with either /e/ or /ə/ (probably because models overfit the data). The substitution /ɔ/ → /o/ is better modeled with a large phoneme window. CRF models are able to map alphabets from canonical to realized. For example, the symbol /ɲ/ in the canonical sequence does not exist in the alphabet used for the realized phoneme annotation. Most models trained with more than 4 min of data are able to adapt the canonical symbol to the realized one. In the context of speech synthesis, some phoneme errors are more harmful than others. For example in Table 4, the substitution /o/ → /ɔ/ or /o/ is not as important as the substitution of /n,j/ → /ʁ/ or /g/. Therefore an adapted error rate based on how close are phonemes could improve statistical approaches for speech synthesis.

6   Conclusion

In this paper, we have presented a pronunciation adaptation method which adapts phonemes generated by the phonetizer to the speech corpus. A CRF pronunciation model trained with linguistic, phonological, articulatory and prosodic features predicts an adapted phoneme sequence from a canonical phoneme sequence. The present work investigates an optimal feature set (features and phoneme window) and a minimal quantity of training material for TTS.

First a cross-validation forward feature selection methodology is proposed. This method allows to select 15 linguistic, phonological and prosodic features (LPrPh). Different feature group combinations are tested together with different phoneme window sizes. An optimal feature set (W2-CLPrPh) brings the best improvement (-8.5 pp) in terms of PER on the validation set. Hence, we have shown that pronunciation adaptation to the speech corpus itself helps to significantly reduce the inconsistency between phonemes as labeled in the underlying speech corpus and those generated by the phonetizer. Moreover, a statistical approach has the advantage of being easily reproducible.

Second, a cross-validation experiment was conducted with decreasing quantities of training material. We can conclude from these experiments that there is a threshold for duration of training data at almost 5 min. Over this threshold, the addition of new data has a high cost but a weak improvement in accuracy: multiplying the duration of training data by 10 improves the PER of 0.5 pp. An ideal PER $= 0$ would be reached for $3 \cdot 10^8$ hours of training data with W2-CLPrPh configuration. Therefore for exploratory researches on pronunciation, 5 min of training data seem to be enough. However, for end-user applications, the more data, the better.

The advantage of large-scaled data is to introduce several weighted phoneme adaptations for each canonical phoneme. Then, including n-best predicted phonemes into phoneme lattices together with weighted phoneme errors could be relevant for speech synthesis applications. Apart from improving TTS quality, the presented pronunciation adaptation method brings interesting perspectives in the use of small-scaled speech corpora for expressive TTS.

Acknowledgments. This study has been realized under the ANR (French National Research Agency) project SynPaFlex ANR-15-CE23-0015.

References

1. Olinsky, C., Cummins, F.: Iterative English adaptation in a speech synthesis system. In: IEEE Workshop on Speech Synthesis (2002)
2. Govind, D., Prasanna, S.M.: Expressive speech synthesis: a review. Int. J. Speech Technol. **16**, 237–260 (2013)
3. Karanasou, P., Yvon, F., Lavergne, T., Lamel, L.: Discriminative training of a phoneme confusion model for a dynamic lexicon in ASR. In: Proceedings of Interspeech (2013)
4. Rao, K., Peng, F., Sak, H., Beaufays, F.: Grapheme-to-phoneme conversion using long short-term memory recurrent neural networks. In: Proceedings of ICASSP (2015)
5. Yao, K., Zweig, G.: Sequence-to-sequence neural net models for grapheme-to-phoneme conversion. In: Proceedings of Interspeech (2015)
6. Lecorvé, G., Lolive, D.: Adaptive statistical utterance phonetization for French. In: Proceedings of ICASSP (2015)
7. Hazen, T.J., Hetherington, I., Shu, H., Livescu, K.: Pronunciation modeling using a finite-state transducer representation. Speech Commun. **46**, 189–203 (2005)
8. Livescu, K., Jyothi, P., Fosler-Lussier, E.: Articulatory feature-based pronunciation modeling. Comput. Speech Lang. **36**, 212–232 (2016)
9. Nagòrski, A., Boves, L., Steeneken, H.: In search of optimal data selection for training of automatic speech recognition systems. In: Proceedings of ASRU (2003)
10. Moore, R.K.: A comparison of the data requirements of automatic speech recognition systems and human listeners. In: Proceedings of Eurospeech (2003)

11. Schuller, B., Batliner, A., Seppi, D., Steidl, S., Vogt, T., Wagner, J., Devillers, L., Vidrascu, L., Amir, N., Kessous, L., Aharonson, V.: The relevance of feature type for the automatic classification of emotional user states: low level descriptors and functionals. In: Proceedings of Interspeech (2007)

12. Tahon, M., Devillers, L.: Towards a small set of robust acoustic features for emotion recognition: challenges. IEEE/ACM Trans. Speech Audio Lang. Process. **54**(1), 16–48 (2016)

13. Chen, Y., Ganapathi, A., Katz, R.: Challenges and opportunities for managing data systems using statistical models. In: Bulletin of the IEEE Computer Society Technical Committee on Data Engineering (2011)

14. Qader, R., Lecorvé, G., Lolive, D., Sébillot, P.: Probabilistic speaker pronunciation adaptation for spontaneous speech synthesis using linguistic features. In: Dediu, A.-H., et al. (eds.) SLSP 2015. LNCS, vol. 9449, pp. 229–241. Springer, Heidelberg (2015). doi:10.1007/978-3-319-25789-1_22

15. Chevelu, J., Lecorvé, G., Lolive, D.: ROOTS: a toolkit for easy, fast and consistent processing of large sequential annotated data collections. In: Proceedings of LREC (2014)

16. Béchet, F.: LIA-PHON: un système complet de phonétisation de texte. Traitement Automatique des Langues (TAL) **42**, 47–67 (2001)

17. Lin, Y., Michel, J.-B., Aiden, E.L., Orwant, J., Brockman, W., Petrov, S.: Syntactic annotations for the Google books ngram corpus. In: Proceedings of ACL (2012)

18. d'Alessandro, C., Rosset, S., Rossi, J.-P.: The pitch of short-duration fundamental frequency glissandos. J. Acoust. Soc. Am. **104**, 2339–2348 (1998)

19. Lavergne, T., Cappé, O., Yvon, F.: Practical very large scale CRFs. In: Proceedings of ACL (2010)

20. Guyon, I., Elissef, A.: An introduction to variable and feature selection. J. Mach. Learn. Res. **3**, 1157–1182 (2003)

21. Tahon, M., Qader, R., Lecorvé, G., Lolive, D.: Improving TTS with corpus-specific pronunciation adaptation. In: Proceedings of Interspeech (2016)

22. Qader, R., Lecorvé, G., Lolive, D., Sébillot, P.: Adaptation de la prononciation pour la synthèse de la parole spontanée en utilisant des informations linguistiques. In: Proceedings of Journées d'Etudes sur la Parole (2016)

A New Perspective on Combining GMM and DNN Frameworks for Speaker Adaptation

Natalia Tomashenko[1,2,3(✉)], Yuri Khokhlov[3], and Yannick Estève[1]

[1] University of Le Mans, Le Mans, France
{natalia.tomashenko,yannick.esteve}@univ-lemans.fr
[2] ITMO University, Saint-Petersburg, Russia
[3] STC-innovations Ltd, Saint-Petersburg, Russia
khokhlov@speechpro.com

Abstract. In this paper we investigate the GMM-derived features for adaptation of context-dependent deep neural network HMM (CD-DNN-HMM) acoustic models with the focus on exploration of fusion of the adapted GMM-derived features and the conventional bottleneck features. We analyze and compare different types of fusion, such as feature level, posterior level, lattice level and others in order to discover the best possible way of fusion. Experimental results on the TED-LIUM corpus show that the proposed adaptation technique can be effectively integrated into DNN setup at different levels and provide additional gain in recognition performance: up to 6 % of relative word error rate reduction (WERR) over the strong speaker adapted DNN baseline, and up to 22 % of relative WERR in comparison with a speaker independent DNN baseline model, trained on conventional features.

Keywords: Speaker adaptation · Deep neural networks (DNN) · MAP · fMLLR · CD-DNN-HMM · GMM-derived (GMMD) features · Fusion · Posterior fusion · Confusion network combination

1 Introduction

Adaptation of DNN acoustic models is a rapidly developing area of research. In the recent years DNNs have replaced conventional Gaussian mixture models (GMM) HMMs in most state-of-the-art automatic speech recognition (ASR) systems, because it has been shown that DNN-HMM models outperform GMM-HMMs in different ASR tasks. However, many adaptation algorithms that have been developed for GMM-HMM systems [1,2] cannot be easily applied to DNNs because of the different nature of these models.

Various adaptation methods have been developed for DNNs. One of the first adaptation methods developed for DNN was *linear transformation* that can be applied at different levels of the DNN-HMM system: to the input features, as in linear input network transformation (LIN) [3,4] or feature-space discriminative linear regression (fDLR) [5,6]; to the activations of hidden layers, as in linear hidden network transformation (LHN) [3]; or to the softmax layer, as in LON

© Springer International Publishing AG 2016
P. Král and C. Martín-Vide (Eds.): SLSP 2016, LNAI 9918, pp. 120–132, 2016.
DOI: 10.1007/978-3-319-45925-7_10

[4] or in output-feature discriminative linear regression [6]. In order to improve generalization during the adaptation *regularization techniques*, such as L2-prior regularization [7], Kullback-Leibler divergence regularization [8], conservative training [9] and others [10] are used. There are also several *model-space adaptation* methods [11–14], such as learning speaker-specific hidden unit contributions (LHUC) [12], the adaptation parameters estimation via maximum a posteriori (MAP) linear regression [13] and hierarchical MAP approach [14]. The concept of *multi-task learning* (MTL) has recently been applied to the task of speaker adaptation in several works [15–17] and has been shown to improve the performance of different model-based DNN adaptation techniques, such as LHN [16] and LHUC [17]. Using *auxiliary features*, such as i-vectors [18–20], is another widely used approach in which the acoustic feature vectors are augmented with additional speaker-specific or channel-specific features computed for each speaker or utterance at both training and test stages. Alternative methods are adaptation with speaker codes [21] and factorized adaptation [22].

However, among the adaptation methods developed for DNNs, only a few take advantage of robust adaptability of GMMs [5,23–28]. The most common way of combining GMM and DNN models for adaptation is using GMM-adapted features, for example fMLLR, as input for DNN training [5,23,24,28]. In [25] likelihood scores from DNN and GMM models, both adapted in the feature space using the same fMLLR transform, are combined at the state level during decoding. Other methods include temporally varying weight regression [26] and GMM-derived (GMMD) features [29–32].

In this paper we continue to investigate GMM framework for adaptation of DNN-HMM acoustic models. Our approach is based on using features derived from a GMM model for training DNN models [29,31,33,34] and GMM-based adaptation techniques. In the previous works it was shown that GMM log-likelihoods can be effectively used as features for training a DNN HMM model, as well as for the speaker adaptation task. In this work we experiment with combination of GMMD features and conventional features at different levels of DNN-HMM based ASR system in order to discover the best possible way of fusion. Also we explore a novel approach for the combination of MAP and fMLLR techniques for SAT with GMMD features.

The rest of the paper is organized as follows. Section 2 describes bottleneck-based GMMD features for adaptation of DNN acoustic models. Section 3 presents several types of combination of GMMD with conventional features at different levels of a DNN architecture. The experimental results are given in Sect. 4. Finally, conclusions are presented in Sect. 5.

2   Bottleneck-Based GMMD Features for Adaptation of DNN Acoustic Models

Construction of GMMD features for adapting DNNs was proposed in [29,31], where it was demonstrated, using MAP and fMLLR adaptation as an example, that this type of features makes it possible to effectively use GMM-HMM adaptation algorithms in the DNN framework. In this section we describe the improved scheme for GMMD feature extraction and application of the concept of GMMD features with adaptation to state-of-the art DNN architecture.

We use bottleneck (BN) features [35] from DNN to train GMM model for GMMD feature extraction. The motivation for using BN features in this approach is that for the better source features we can obtain better adaptation results. BN features allow us to capture long term spectro-temporal dynamics of the signal with GMDD features and are proven to be effective both for GMM and DNN acoustic model training [35].

The scheme for training DNN model with GMM adaptation framework is shown in Fig. 1. First, 40-dimensional log-scale filterbank features concatenated with 3-dimensional pitch-features, are spliced across 11 neighboring frames (5 frames on each side of the current frame), resulting in 473-dimensional (43×11) feature vectors. After that a DCT transform is applied and the dimension is reduced to 258. Then a DNN model for 40-dimensional BN features is trained on these features. An auxiliary triphone or monophone GMM model is used to transform BN feature vectors into log-likelihoods vectors. At this step, speaker adaptation of the auxiliary SI GMM-HMM model is performed for each speaker in the training corpus and a new speaker-adapted (SA) GMM-HMM model is created in order to obtain SA GMMD features.

For a given BN feature vector, a new GMM-derived feature vector is obtained by calculating log-likelihoods across all the states of the auxiliary GMM model on the given vector. Suppose o_t is the BN feature vector at time t, then the new GMM-derived feature vector f_t is calculated as follows:

$$f_t = [p_t^1, \ldots, p_t^n], \tag{1}$$

where n is the number of states in the auxiliary GMM-HMM model,

$$p_t^i = \log\left(P(o_t \mid s_t = i)\right) \tag{2}$$

is the log-likelihood estimated using the GMM-HMM. Here s_t denotes the state index at time t. After that, the features are spliced in time taking a context size of 13 frames: $[-10, -5..5, 10]$. We will refer to these resulting features as GMMD features. These features are used as the input for training the DNN. The proposed approach can be considered a feature space transformation technique with respect to DNN-HMMs trained on GMMD features.

3   System Fusion

In this section we suggest several types of combination of GMM-derived features with conventional ones at different levels of DNN architecture. It is known that GMM and DNN models can be complementary and their combination allows to improve the performance of ASR systems [34,36]. Fusion is useful when the individual systems, used in combination, contain complementary information. To obtain better recognition performance we explore the following types of fusion.

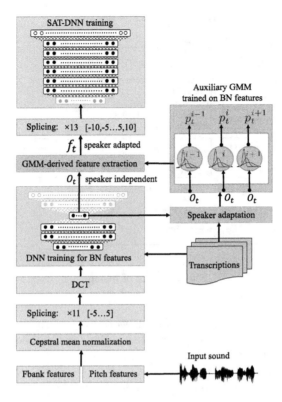

Fig. 1. Using speaker adapted BN-based GMM-derived features for SAT DNN-HMM training.

3.1 Feature Level Fusion

In this type of fusion, also called *early fusion* or *early integration* [34], input features are combined before performing classification, as shown in Fig. 2a. In our case features of different types - GMMD features and cepstral or BN features are simply concatenated and inputted into the DNN model for training. This type of fusion allows to combine different adaptation techniques in a one DNN model. For example, MAP-adapted GMMD features can be concatenated with fMLLR-adapted BN features, that makes adaptation more efficient for both small and large adaptation sets.

3.2 Posterior Level Fusion

Posterior level fusion is also referred to as *late fusion* [28], *late integration* [34], *state-level score combination* [25] or *explicit combination* [36]. In this type of fusion (Fig. 2b) the outputs of two or more DNN models are combined on a state level. The outputs of the two classifiers can be combined using various multi-stream combination techniques such as sum (or linear combination),

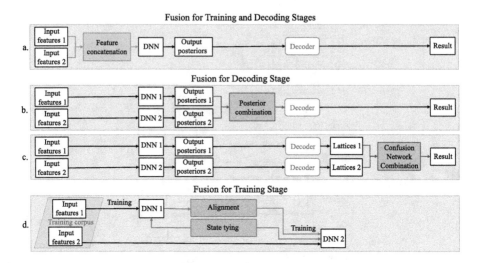

Fig. 2. Types of fusion

product, maximum, minimum, etc. In this work we perform frame-synchronous fusion using a linear combination of the observation log-likelihoods of two models (DNN_1 and DNN_2) as follows:

$$\log\left(p(o_t \mid s_i)\right) = \alpha \log(p_{\mathrm{DNN}_1}(o_t \mid s_i)) + (1 - \alpha)\log(p_{\mathrm{DNN}_2}(o_t \mid s_i)), \qquad (3)$$

where $\alpha \in [0, 1]$ is a weight factor that is optimized on a development set. This approach assumes that both models have the same state tying structure.

3.3   Lattice Level Fusion

The highest level of fusion operates in the space of generated word hypothesis and tries to rescore or modify recognition hypotheses provided as lattices (Fig. 2c) or n-best lists. This type of fusion is also referred to as *implicit combination* [36].

One of the most common techniques for ASR system combination are Recognizer Output Voting Error Reduction (ROVER) [37] and Confusion Network Combination (CNC) [38]. In ROVER 1-best word sequences from the different ASR systems are combined into a single word transition network (WTN) using a dynamic programming algorithm. Based on this WTN the best scoring word is chosen among the words aligned together. The decision is based either on the voting scheme, or word confidence scores if they are available for all systems. In CNC instead of aligning the single best output, confusion networks, built from individual lattices, are aligned. In this work we experiment with CNC approach because usually (for example, [38]) it provides better results than a simple ROVER scheme.

3.4   Multiple and Other Types of Fusion

There are other possible ways of combining information from different ASR systems, than those listed above, that also could be considered as fusion. In this paper we implement the two following approaches.

The first approach is specific to the adaptation task and is related only to the acoustic model adaptation stage. It consists in using the transcripts (or lattices) obtained from the decoding pass of the one ASR system in order to adapt another ASR system.

The second approach (Fig. 2d) is related to the acoustic model training. It is possible to transfer some information from building one acoustic model to another one. In this paper we use phoneme-to-speech alignment obtained by one acoustic DNN model to train another DNN model. In addition we use state tying from the first DNN model to train the second DNN. This procedure is important when we want to apply *posterior fusion* for two DNNs and need the same state tying for these models.

By *multiple fusion* we mean a combination of several types of fusion described above.

4   Experimental Results

4.1   Data Sets

The experiments were conducted on the TED-LIUM corpus [39]. We used the last (second) release of this corpus. This publicly available data set contains 1495 TED talks that amount to 207 h (141 h of male, 66 h of female) speech data from 1242 speakers, 16 kHz. For experiments with SAT and adaptation we removed from the original corpus data for those speakers, who had less than 5 min of data, and from the rest of the corpus we made four data sets: training set, development set and two test sets. Characteristics of the obtained data sets are given in Table 1. The motivation for creating the new test and development data sets was to obtain data sets, that are more representative and balanced in characteristics (gender, duration) than the original ones and more suitable for adaptation experiments.

For evaluation we use 150K word vocabulary and publicly available trigram language model *cantab-TEDLIUM-pruned.lm3*[1].

4.2   Baseline System

We used the open-source Kaldi toolkit [40] and followed mostly the standard TED-LIUM Kaldi recipe to train the baseline system.

For training DNN models, first the initial GMM model was trained using 39-dimensional MFCC features with delta and acceleration coefficients. Linear discriminant analysis (LDA) followed by maximum likelihood linear transform

---

[1] http://cantabresearch.com/cantab-TEDLIUM.tar.bz2.

Table 1. Data sets statistics

Characteristic		Data set			
		Training	Development	Test$_1$	Test$_2$
Duration, hours	Total	171.66	3.49	3.49	4.90
	Male	120.50	1.76	1.76	3.51
	Female	51.15	1.73	1.73	1.39
Duration per speaker, minutes	Mean	10.0	15.0	15.0	21.0
	Minimum	5.0	14.4	14.4	18.3
	Maximum	18.3	15.4	15.4	24.9
Number of speakers	Total	1029	14	14	14
	Male	710	7	7	10
	Female	319	7	7	4
Number of words	Total	-	36672	35555	51452

(MLLT) and fMLLR transformation was then applied over these MFCC features to build a GMM-HMM system. Discriminative training with the boosted maximum mutual information (BMMI) objective was finally performed on top of this model.

SAT DNN on fMLLR Features. Then a DNN was trained for BN feature extraction. The DNN system was trained using the frame-level cross entropy criterion and the senone alignment generated from the GMM system. For training this DNN, 40-dimensional log-scale filterbank features concatenated with 3-dimensional pitch-features, were spliced across 11 neighbouring frames, resulting in 473-dimensional (43×11) feature vectors. After that a DCT transform was applied and the dimension was reduced to 258. A DNN model for extraction 40-dimensional BN features was trained with the following topology: 258-dimensional input layer; four hidden layers (HL), where the third HL was a BN layer with 40 neurons and other three HLs were 1500-dimensional; the output layer was 2390-dimensional. On the obtained BN features we trained the GMM model, which was used to produce the forced alignment, and then SAT-GMM model was trained on fMLLR-adapted BN features. Then for training the final DNN model, fMLLR-adapted BN features were spliced in time with the context of 13 frames: $[-10, -5..5, 10]$. The final DNN had 520-dimensional input layer; six 2048-dimensional HLs with logistic sigmoid activation function, and 4184-dimensional softmax output layer, with units corresponding to the CD states. The DNN parameters were initialized with stacked restricted Boltzmann machines (RBMs) by using layer by layer generative pre-training. It was trained with an initial learning rate of 0.008 using the cross-entropy objective function. After that four epochs of sequence-discriminative training with per-utterance updates, optimizing state Minimum Bayes Risk (sMBR) criteria, were performed.

Baseline SI-DNN. This model was trained in a similar way as the SAT DNN described above, but without fMLLR adaptation.

4.3   Adaptation and Fusion Results

The adaptation experiments were conducted in an unsupervised mode on the test data using the transcripts obtained from the first decoding pass obtained by the baseline SAT-DNN model. We empirically studied different types of fusion described in Sect. 3 and applied them to DNN models trained using GMMD-features, extracted as proposed in Sect. 2. The performance results in terms of WER for SI and SAT DNN-HMM models are presented in Table 2. The first two lines of the table correspond to the baseline SI (#1) and SAT (#2) DNNs, which were trained as described in Sect. 4.2.

For investigating feature level fusion (Sect. 3.1) we trained two SAT DNN acoustic models with different schemes for feature extraction, as shown in Figs. 3 and 4. Features AF_1 (Fig. 3) were extracted using monophone auxilluary GMM model, trained on BN features. For this type of features we trained two DNN models. For training the first model (#3) we took state tying from the baseline SAT-DNN. The purpose of using the same state tying is to allow posterior level fusion for these models. For training the second model (#4) on AF_1 features, in addition to the state tying structure, we also took an alignment, obtained by SAT-DNN model (#2).

Features AF_2 (Fig. 4) were extracted using a triphone auxilluary GMM model, trained on the same BN features. Using triphone GMMs allows to adapt more classes, and can lead to more precise adaptation, when the amount of the adaptation data increases. However this change leads to the increasing of GMMD feature dimension. Hence, we have to reduce the feature dimension of GMMD features before splicing and concatenating with BN features. There are many possible ways for solving this problem, such as principal components analysis (PCA), Linear Discriminant Analysis (LDA) or Heteroscedastic LDA (HLDA) and others. In this work we used the concept of BN features for this purpose. First, we spliced GMMD features with context: [−5...5], and then trained 40-dimensional BN features (with the DNN topology, which was similar to the one described in Sect. 4.2). Then we concatenated two types of BN features and spliced them for training the final DNN model (#5). All three DNN models were trained on the proposed features in the same manner and had the same topology except for the input features, as the final baseline SAT DNN model, trained on BN features (Sect. 4.2). For comparison purpose with lattice-based fusion (#9–11) we report WER of the consensus hypothesis in parentheses for all other experiments (#2–8).

Parameter τ in Table 2 is the parameter in MAP adaptation, that controls the balance between the maximum likelihood estimate of the mean and its prior value [2,29]. For acoustic model training τ was set equal to 5. For testing we tried different τ and optimized it on the development set.

After that we made posterior fusion of the obtained three models (#3, 4, 5) and the baseline SAT-DNN model (#2). Results are given in lines #6, 7, 8

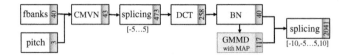

Fig. 3. Adapted features AF_1: for monophone auxiliary GMM model

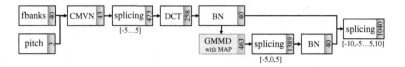

Fig. 4. Adapted features AF_2: for triphone auxiliary GMM model

Table 2. Summary of the adaptation results for DNN models. The results in parentheses correspond to WER of the consensus hypothesis. Here ST denotes state tying, and $\downarrow$ – relative WERR (for consensus hypothesis) in comparison with AM trained on fMLLR-BN (#2). The bold figures in the table indicate the best performance improvement.

#	Features	τ	WER,%		
			Development	Test$_1$	Test$_2$
1	SI-BN		12.14	10.77	13.75
2	fMLLR-BN		10.64 (10.57)	9.52 (9.46)	12.78 (12.67)
3	AF_1, ST #2	2	10.27 (10.21)	9.59 (9.56)	12.94 (12.88)
4	AF_1, ST and align. #2	5	10.26 (10.23)	9.40 (9.31)	12.52 (12.46)
5	AF_2, ST and align. #2	5	10.42 (10.37)	9.74 (9.69)	13.29 (13.23)
Posterior fusion					
6	#2 and #3, $\alpha = 0.55$	2	10.09 (10.01) $\downarrow$ 5.3	9.12 (9.03) $\downarrow$ **4.6**	12.23 (12.19) $\downarrow$ 3.8
7	#2 and #4, $\alpha = 0.55$	5	9.98 (9.91) $\downarrow$ **6.2**	9.15 (9.06) $\downarrow$ 4.3	12.11 (12.04) $\downarrow$ **5.0**
8	#2 and #5, $\alpha = 0.45$	5	9.96 (9.91) $\downarrow$ **6.2**	9.12 (9.10) $\downarrow$ 3.8	12.36 (12.23) $\downarrow$ 3.5
Lattice fusion					
9	#2 and #3, $\alpha = 0.54$	2	9.93 $\downarrow$ 6.0	9.20 $\downarrow$ 2.8	12.31 $\downarrow$ 2.9
10	#2 and #4, $\alpha = 0.56$	5	10.06 $\downarrow$ 4.8	9.09 $\downarrow$ 4.0	12.12 $\downarrow$ 4.4
11	#2 and #5, $\alpha = 0.50$	5	10.01 $\downarrow$ 5.3	9.17 $\downarrow$ 3.1	12.25 $\downarrow$ 3.3

correspondingly. Value $(1 - \alpha)$ in Formula 3 is a weight of the baseline SAT-DNN model. Parameter α was optimized on the development set.

Finally we applied lattice fusion for the same pairs of models (Lines #9, 10, 11). In this type of fusion, before merging lattices, for each edge, scores were replaced by its a posteriori probabilities. Posteriors were computed for each lattice independently. The optimal normalizing factors for each model were found independently on the development set. Then the two lattices were merged into a single lattice and posteriors were weighted using parameter α. As before, value $(1 - \alpha)$ in Formula 3 corresponds to the baseline SAT-DNN model. The resulting lattice was converted into the CN and the final result was obtained from this CN.

We can see, that both - posterior and lattice types of fusion provide similar improvement for all three models: approximately 3 %–6 % of relative WER reduction (WERR) in comparison with the adapted baseline model (SAT DNN on fMLLR features, #2), and 11 %–22 % of relative WERR in comparison with the SI baseline model (#1). For models #3–5 only MAP adaptation was applied. Experiments #6–11 present combination of two different adaptation types: MAP and fMLLR. Is this interesting to note that in all experiments optimal value of α is close to 0.5, so all types of models are equally important for fusion. We can see that MAP adaptation on GMMD features can be complementary to fMLLR adaptation on conventional BN features.

5   Conclusions

In this paper we have investigated the GMM framework for adaptation of DNN-HMM acoustic models and combination of GMM-derived with conventional features at different levels of DNN architecture. Experimental results on the TED-LIUM corpus demonstrate that, in an unsupervised adaptation mode, the proposed adaptation and fusion techniques can provide approximately, a 11–22 % relative WERR on different adaptation sets, compared to the SI DNN system built on conventional features, and a 3–6 % relative WERR compared to the strong adapted baseline - SAT-DNN trained on fMLLR adapted features. Experiments with different types of fusion show that MAP adaptation on GMMD features can be complementary to fMLLR adaptation on conventional BN features. We can see, that two types of fusion – posterior level and lattice level provide additional comparable improvement, and in most cases posterior level fusion provides slightly better results than the lattice level fusion. As for the feature level combination, we can conclude, that on average, slightly better results are archived for model #4.

Acknowledgements. This work was partially funded by the European Commission through the EUMSSI project, under the contract number 611057, in the framework of the FP7-ICT-2013-10 call and by the Government of the Russian Federation, Grant 074-U01.

References

1. Gales, M.J.: Maximum likelihood linear transformations for HMM-based speech recognition. Comput. Speech Lang. **12**(2), 75–98 (1998)
2. Gauvain, J.-L., Lee, C.-H.: Maximum a posteriori estimation for multivariate Gaussian mixture observations of Markov chains. IEEE Trans. Speech Audio Process. **2**, 291–298 (1994)
3. Gemello, R., Mana, F., Scanzio, S., Laface, P., De Mori, R.: Adaptation of hybrid ANN/HMM models using linear hidden transformations and conservative training. In: Proceedings of ICASSP (2006)
4. Li, B., Sim, K.C.: Comparison of discriminative input and output transformations for speaker adaptation in the hybrid NN/HMM systems, pp. 526–529 (2010)
5. Seide, F., Li, G., Chen, X., Yu, D.: Feature engineering in context-dependent deep neural networks for conversational speech transcription. In: Proceedings of ASRU, pp. 24–29. IEEE (2011)
6. Yao, K., Yu, D., Seide, F., Su, H., Deng, L., Gong, Y.: Adaptation of context-dependent deep neural networks for automatic speech recognition. In: Proceedings of SLT, pp. 366–369. IEEE (2012)
7. Liao, H.: Speaker adaptation of context dependent deep neural networks. In: Proceedings of ICASSP, pp. 7947–7951. IEEE (2013)
8. Yu, D., Yao, K., Su, H., Li, G., Seide, F.: KL-divergence regularized deep neural network adaptation for improved large vocabulary speech recognition. In: Proceedings of ICASSP, pp. 7893–7897 (2013)
9. Albesano, D., Gemello, R., Laface, P., Mana, F., Scanzio, S.: Adaptation of artificial neural networks avoiding catastrophic forgetting. In: Proceedings of IJCNN 2006, pp. 1554–1561. IEEE (2006)
10. Ochiai, T., Matsuda, S., Lu, X., Hori, C., Katagiri, S.: Speaker adaptive training using deep neural networks. In: Proceedings of ICASSP, pp. 6349–6353. IEEE (2014)
11. Siniscalchi, S.M., Li, J., Lee, C.-H.: Hermitian polynomial for speaker adaptation of connectionist speech recognition systems. IEEE Trans. Audio Speech Lang. Process. **21**(10), 2152–2161 (2013)
12. Swietojanski, P., Renals, S.: Learning hidden unit contributions for unsupervised speaker adaptation of neural network acoustic models. In: Proceedings of SLT, pp. 171–176. IEEE (2014)
13. Huang, Z., Li, J., Siniscalchi, S.M., Chen, I.-F., Weng, C., Lee, C.-H.: Feature space maximum a posteriori linear regression for adaptation of deep neural networks. In: Proceedings of INTERSPEECH, pp. 2992–2996 (2014)
14. Huang, Z., Siniscalchi, S.M., Chen, I.-F., Li, J., Wu, J., Lee, C.-H.: Maximum a posteriori adaptation of network parameters in deep models. In: Proceedings of INTERSPEECH (2015)
15. Li, S., Lu, X., Akita, Y., Kawahara, T.: Ensemble speaker modeling using speaker adaptive training deep neural network for speaker adaptation. In: Proceedings of INTERSPEECH (2015)
16. Huang, Z., Li, J., Siniscalchi, S.M., Chen, I.-F., Wu, J., Lee, C.-H.: Rapid adaptation for deep neural networks through multi-task learning. In: Proceedings of INTERSPEECH (2015)
17. Swietojanski, P., Bell, P., Renals, S.: Structured output layer with auxiliary targets for context-dependent acoustic modelling. In: Proceedings of INTERSPEECH (2015)

18. Karanasou, P., Wang, Y., Gales, M.J., Woodland, P.C.: Adaptation of deep neural network acoustic models using factorised i-vectors. In: Proceedings of INTERSPEECH, pp. 2180–2184 (2014)
19. Gupta, V., Kenny, P., Ouellet, P., Stafylakis, T.: I-vector-based speaker adaptation of deep neural networks for french broadcast audio transcription. In: Proceedings of ICASSP, pp. 6334–6338. IEEE (2014)
20. Senior, A., Lopez-Moreno, I.: Improving DNN speaker independence with i-vector inputs. In: Proceedings of ICASSP, pp. 225–229 (2014)
21. Xue, S., Abdel-Hamid, O., Jiang, H., Dai, L., Liu, Q.: Fast adaptation of deep neural network based on discriminant codes for speech recognition. IEEE/ACM Trans. Audio Speech Lang. Process. **22**(12), 1713–1725 (2014)
22. Li, J., Huang, J.-T., Gong, Y.: Factorized adaptation for deep neural network. In: Proceedings of ICASSP, pp. 5537–5541. IEEE (2014)
23. Rath, S.P., Povey, D., Veselý, K., Cernocký, J.: Improved feature processing for deep neural networks. In: Proceedings of INTERSPEECH, pp. 109–113 (2013)
24. Kanagawa, H., Tachioka, Y., Watanabe, S., Ishii, J.: Feature-space structural MAPLR with regression tree-based multiple transformation matrices for DNN (2015)
25. Lei, X., Lin, H., Heigold, G.: Deep neural networks with auxiliary Gaussian mixture models for real-time speech recognition. In: Proceedings of ICASSP, pp. 7634–7638. IEEE (2013)
26. Liu, S., Sim, K.C.: On combining DNN and GMM with unsupervised speaker adaptation for robust automatic speech recognition. In: Proceedings of ICASSP, pp. 195–199. IEEE (2014)
27. Murali Karthick, B., Kolhar, P., Umesh, S.: Speaker adaptation of convolutional neural network using speaker specific subspace vectors of SGMM (2015)
28. Parthasarathi, S.H.K., Hoffmeister, B., Matsoukas, S., Mandal, A., Strom, N., Garimella, S.: fMLLR based feature-space speaker adaptation of DNN acoustic models. In: Proceedings of INTERSPEECH (2015)
29. Tomashenko, N., Khokhlov, Y.: Speaker adaptation of context dependent deep neural networks based on MAP-adaptation and GMM-derived feature processing. In: Proceedings of INTERSPEECH, pp. 2997–3001 (2014)
30. Tomashenko, N., Khokhlov, Y., Larcher, A., Estève, Y.: Exploring GMM-derived features for unsupervised adaptation of deep neural network acoustic models. In: Ronzhin, A., Potapova, R., Németh, G. (eds.) SPECOM 2016. LNCS, vol. 9811, pp. 304–311. Springer, Heidelberg (2016). doi:10.1007/978-3-319-43958-7_36
31. Tomashenko, N., Khokhlov, Y.: GMM-derived features for effective unsupervised adaptation of deep neural network acoustic models. In: Proceedings of INTERSPEECH, pp. 2882–2886 (2015)
32. Tomashenko, N., Khokhlov, Y., Larcher, A., Estève, Y.: Exploration de paramètres acoustiques dérivés de GMM pour l'adaptation non supervisée de modèles acoustiques à base de réseaux de neurones profonds. In: Proceedings of 31éme Journées d'Études sur la Parole (JEP), pp. 337–345 (2016)
33. Tomashenko, N., Khokhlov, Y., Esteve, Y.: On the use of Gaussian mixture model framework to improve speaker adaptation of deep neural network acoustic models. In: Proceedings of INTERSPEECH (2016)
34. Pinto, J.P., Hermansky, H.: Combining evidence from a generative and a discriminative model in phoneme recognition. Technical report, IDIAP (2008)
35. Grézl, F., Karafiát, M., Vesely, K.: Adaptation of multilingual stacked bottle-neck neural network structure for new language. In: Proceedings of ICASSP, pp. 7654–7658 (2014)

36. Swietojanski, P., Ghoshal, A., Renals, S.: Revisiting hybrid and GMM-HMM system combination techniques. In: Proceedings of ICASSP, pp. 6744–6748. IEEE (2013)

37. Fiscus, J.G.: A post-processing system to yield reduced word error rates: recognizer output voting error reduction (ROVER). In: Proceedings of ASRU, pp. 347–354. IEEE (1997)

38. Evermann, G., Woodland, P.: Posterior probability decoding, confidence estimation and system combination. In: Proceedings of Speech Transcription Workshop, Baltimore, vol. 27 (2000)

39. Rousseau, A., Deléglise, P., Estève, Y.: Enhancing the TED-LIUM corpus with selected data for language modeling and more TED talks. In: Proceedings of LREC, pp. 3935–3939 (2014)

40. Povey, D., Ghoshal, A., Boulianne, G., Burget, L., Glembek, O., Goel, N., Hannemann, M., Motlicek, P., Qian, Y., Schwarz, P., et al.: The Kaldi speech recognition toolkit. In: Proceedings of ASRU (2011)

Class n-Gram Models for Very Large Vocabulary Speech Recognition of Finnish and Estonian

Matti Varjokallio[1]([✉]), Mikko Kurimo[1], and Sami Virpioja[2]

[1] Department of Signal Processing and Acoustics, School of Electrical Engineering,
Aalto University, Espoo, Finland
{matti.varjokallio,mikko.kurimo}@aalto.fi
[2] Department of Computer Science, School of Science, Aalto University,
Espoo, Finland
sami.virpioja@aalto.fi

Abstract. We study class n-gram models for very large vocabulary speech recognition of Finnish and Estonian. The models are trained with vocabulary sizes of several millions of words using automatically derived classes. To evaluate the models on Finnish and an Estonian broadcast news speech recognition task, we modify Aalto University's LVCSR decoder to operate with the class n-grams and very large vocabularies. Linear interpolation of a standard n-gram model and a class n-gram model provides relative perplexity improvements of 21.3 % for Finnish and 12.8 % for Estonian over the n-gram model. The relative improvements in word error rates are 5.5 % for Finnish and 7.4 % for Estonian. We also compare our word-based models to a state-of-the-art unlimited vocabulary recognizer utilizing subword n-gram models, and show that the very large vocabulary word-based models can perform equally well or better.

Keywords: Language modelling · Class n-gram models · Morphologically rich languages · Speech recognition

1 Introduction

The standard solution for language modelling in large vocabulary continuous speech recognition is a statistical n-gram model trained over words. The frequency estimates are smoothed in order to be able to assign probabilities to word sequences not present in the training corpus [6]. For morphologically rich languages, however, the word-based approach is not without shortcomings. As a very large vocabulary is required to achieve a sufficiently small out-of-vocabulary (OOV) rate, even large text databases become sparse for training accurate n-gram models. This manifests itself in the form of low n-gram hit rates and increased error rates.

For agglutinative languages, building the n-gram models over subword units such as statistical morphs has proven to be a solid choice [8]. This way, probabilities may be assigned to word forms which are not necessarily covered by

© Springer International Publishing AG 2016
P. Král and C. Martín-Vide (Eds.): SLSP 2016, LNAI 9918, pp. 133–144, 2016.
DOI: 10.1007/978-3-319-45925-7_11

the training corpus. With subword models, it is possible to opt for either unlimited vocabulary speech recognition [10] or use a fixed vocabulary, but still keep the option of adding new words to the vocabulary if needed [35]. In some cases subwords also provide better n-gram estimates with the same vocabulary [17].

Recent studies for Hungarian [31] and Finnish [35] have, however, shown that carefully implemented word-based n-grams can produce competitive error rates compared to the subword approach. This requires an ASR decoder that is capable of effectively handling a vocabulary of millions of word forms and large n-gram models. In addition, a large training corpus is needed for sufficient coverage of word forms and robust n-gram estimates.

A traditional approach for alleviating the data sparsity issues are the class n-gram models [4,15]. In an early work [23], variable-length category n-grams over part-of-speech tags were trained and evaluated in English speech recognition. Using automatically derived classes and thus increasing the number of classes was found to give larger improvements when interpolated with word n-grams [22]. For morphologically richer languages, class n-grams trained over automatically derived classes have been found to improve language modelling for Russian [38]. In a study on Lithuanian language modelling [34], up to 13 % perplexity reductions were reached by a linear interpolation with a class n-gram using automatically derived classes, while selection of morphologically motivated classes did not improve perplexity. However, in a more recent study on Czech and Slovak language modelling [5], linear interpolation with morphological class n-grams improved perplexities by around 10 % for a large corpus.

In this work, we study class-based language modelling for Finnish and Estonian speech recognition with very large vocabulary sizes. As the size of the vocabulary grows, the importance of the word clustering methods could be expected to increase. Despite the potential of the class n-gram models for the speech recognition of morphologically rich languages, there haven't been many studies on this topic. Class n-gram models have been evaluated for instance in English speech recognition tasks in [18,37]. Some results on Lithuanian speech recognition have been mentioned in [33].

The most common approach for Finnish and Estonian language modelling is to train the language models over statistical morphs or other subword lexical units. As the perplexity values for the subword-based language models and the word-based language models are not directly comparable due to the different OOV rates, their performance needs to be evaluated in a speech recognition task. Due to recent improvements in the decoder design [35], we are able to compare subword language models to word-based language models with a very large vocabulary size. By the linear interpolation of an n-gram model and a class n-gram model, we obtain equal or better results than with an unlimited vocabulary subword recognizer. Stand-alone class n-gram models provide also reasonable error rates and improved robustness with compact model sizes. We are not aware of any earlier work comparing word-based class n-gram interpolated recognizer to a state-of-the-art subword-based unlimited vocabulary recognizer.

2   Methods

2.1   Class n-Grams

The most common form of a class n-gram model [4,15] may be defined as

$$P(w_i|w_{i-(n-1)}^{i-1}) = P(w_i|c_i) * P(c_i|c_{i-(n-1)}^{i-1}), \tag{1}$$

where the words w are clustered into equivalence classes c. The word history is denoted by $w_{i-(n-1)}^{i-1}$ and the corresponding class history by $c_{i-(n-1)}^{i-1}$. After the classification, the class membership probabilities $P(w_i|c_i)$ and the class n-gram component $P(c_i|c_{i-(n-1)}^{i-1})$ are typically estimated as given by the maximum likelihood estimates:

$$P(w|c) = \frac{f(w)}{\sum_{v \in C(w)} f(v)} \tag{2}$$

$$P(c_i|c_{i-(n-1)}^{i-1}) = \frac{f(c_{i-(n-1)}, .., c_i)}{f(c_{i-(n-1)}, .., c_{i-1})}, \tag{3}$$

where $f(w)$ denotes the frequency of the word w and $f(c_{i-(n-1)}, .., c_i)$ the frequency of a class sequence.

Exchange Algorithm. The so-called exchange algorithm for forming statistical word classes with bigram statistics was given in [15]:

Algorithm 1. Exchange algorithm

---
1   compute initial class mapping
2   sum initial class based counts
3   compute initial perplexity
4   **repeat**
5   **foreach** *word w of the vocabulary* **do**
6   remove word from its class
7   **foreach** *class k* **do**
8   tentatively move word w to class k
9   compute perplexity for this exchange
10   move word w to class k with minimum perplexity
11 **until** *stopping criterion is met;*

The algorithm operates by iterating over all the words, evaluating all possible class exchanges for each word, and choosing the exchange that provides the largest improvement for the likelihood. Later work discussed efficient implementations using the word-class and class-word statistics as well as extension to trigram clustering [18]. While trigram statistics may provide improvements for a small number of classes, they often result in overlearning, and the best performance is normally obtained with bigram clustering [3,18]. The evaluation step may be parallelized for each word [3].

Morphologically Motivated Classes. Several studies have experimented with part-of-speech or morphologically motivated classes [5,22,34]. For Finnish, there exists an open source morphological analyzer, Omorfi [25]. Due to the rich morphology of Finnish and fine-grained output of the Omorfi analyzer, the number of different morphological analyses is in the range of 5000–7000. The significantly higher amount of morphological classes compared to the pure part-of-speech categories is promising for language modelling. However, to prevent the increase of the OOV rate of the language model, we need to tag also those word forms not recognized by the analyzer. We tested two different approaches to find out whether the morphological classes could provide results comparable to the purely statistical word clustering approaches.

In the first approach, we used Finnpos [28], an accurate morphological tagger and lemmatizer using conditional random fields, trained from the Omorfi analyses. The training corpus was tagged with Finnpos and the class n-gram model was trained using the resulting morphological class sequence. This approach increases the vocabulary size, as many words have multiple analyses due to the ambiguity. Some improvements were obtained, but the morphological classes were not as efficient as the classes derived by the exchange algorithm.

We also tried a more statistically oriented approach with the Omorfi analyses. Category n-gram [23] is a generalization of the class n-gram model that allows multiple categories per word. It can be shown that one form of a category n-gram model may be trained using the expectation-maximization algorithm. This kind of model performs the tagging statistically and models the morphological disambiguation with alternative categories. In this case the vocabulary size does not increase. In the final steps of the training, the number of categories per word could be reduced to one with only a minor loss in perplexity. At this stage this approach provided better word error rates than the Finnpos approach. These classes were further refined using class merging and splitting, resulting in much improved perplexity values. However, we also found out that the resulting clustering could be improved by running the exchange algorithm on top of these classes. Thus the approach was essentially reduced to an initialization for the exchange algorithm. The perplexity of the language model did not improve compared to a simple initialization where words were assigned to classes by their frequency order. However, there were still differences between the resulting classifications, as the interpolation of the models trained using both initializations gave a further 3.0 % relative improvement in the perplexity.

So far, we have been unable to utilize morphological information in a way that would improve results over the exchange algorithm. The morphological classes may still be useful with less data or for estimating probabilities for out-of-vocabulary words if their grammatical analysis is available. In Sect. 3, we report results for the models that use only the exchange algorithm.

2.2   Subword n-Grams

A popular approach for tackling the OOV and the data sparsity problems for agglutinative languages has been to train the statistical language models over morphs or other subword units. By combining the subword units of the lexicon it is possible to assign probabilities to word forms that do not occur in the training corpus. If the lexicon includes for example all individual letters or syllables of the language, the vocabulary of the recognizer is unlimited [10]. A higher order n-gram model is required to get the full benefit from the subword modelling [13].

Statistical approaches for learning the units have given good results on many languages [8,31]. A popular method is the Morfessor Baseline algorithm [7], which uses the minimum description length (MDL) criterion to find a balance between the cost of storing the model and encoding the training corpus with the model. Morfessor Baseline encodes the corpus with a unigram model.

An alternative method for finding a subword vocabulary is to maximize unigram likelihood via multigram expectation-maximization training [9]. Combined with efficient greedy likelihood-based pruning, it provides lexicons that work well for speech recognition [36]. This approach is used in our work.

In subword-based speech recognition, the word boundaries need to be modelled explicitly. In this work, we use a special word boundary symbol between the words.

2.3   Decoding

Speech recognition decoders can broadly be categorized into static and dynamic decoders [2]. In a static decoder, all data sources are included in the search network, whereas in a dynamic decoder the language model probabilities are applied separately during the decoding. The most common type of a static decoder is based on the use of the weighted finite state transducers (WFST) [20]. The most typical dynamic decoder codes the recognition vocabulary using a lexical prefix tree [21] and performs the search using the token-passing procedure [39]. In this work, we follow the dynamic decoding approach. Standard techniques required for the decoding include the beam search, hypothesis recombination, language model look-ahead [24] and the cross-word modelling [29]. An important property for this work is that large and long-span n-gram models may be efficiently applied with a dynamic decoder. The interpolation with the class n-gram models is also relatively straightforward to do in the first decoding pass.

In this work, we use a modified version of the decoder in the AaltoASR package [1]. The decoder was initially developed mainly for the unlimited vocabulary morph-based recognition task [11,26]. The recognition graph for the subword decoding needs a special construction to correctly handle the intra-word and the inter-word unit boundaries and to allow cross-word pronunciation modelling. The decoder is also able to handle long-span n-gram models [13]. According to an error analysis, only a small part of the recognition errors originate from the search [12].

Even though the word-based recognition is almost by definition a simpler task than the unlimited vocabulary recognition, it has some practical challenges. Even with minimization, the graph size will be large, increasing different bookkeeping costs. Also the look-ahead model is very important for the recognition accuracy, because the word labels are more unevenly distributed in the graph. Recent studies have shown that very large vocabularies may be efficiently decoded using large n-gram models [30,35].

As the perplexities for the word-based and the subword-based models are not directly comparable due to the different OOV rates, we compare their performance in a speech recognition task. The same recognizer implementation is applied for both models, but the recognition graph is constructed differently. Silence and cross-word modelling in the graphs are identical. An important operation in the decoding is the so-called hypothesis recombination. If there are several tokens in the same graph node and in the same n-gram model state, only the best token is kept and the rest discarded. The hypothesis recombination is extended for the class n-gram interpolation by applying the recombination on n-gram and class n-gram state tuples. This way the class n-grams are applied without additional approximations to the beam search. Following [35], we use a unigram look-ahead model with the word n-grams and a bigram look-ahead model with the subword n-grams.

3   Experiments

3.1   Experimental Setup

For training the Finnish language models, we used the CSC Kielipankki corpus [32]. The corpus contains text from Finnish newspapers, magazines and books. The size of the corpus is around 150M word tokens and 4.1M word types. The list of words was filtered by simple phonotactic filters and by cross-checking the singleton words using counts from a web corpus. The vocabulary size was limited to 2.4M word types, with only a small impact on the recognition accuracy. Also discarding the singleton words would be possible. Attention was paid not to optimize the vocabulary in any specific way for the recognition task.

Estonian language models were trained on a corpus of Estonian newspaper articles and news articles from the web [19]. The training corpus consisted of 80M word tokens with 1.6M distinct word types. All the word types were included in the vocabulary.

The Finnish acoustic models were trained with audio from the Speecon database [14]. A 31-h set of clean dictated wideband speech from 310 speakers was used for training. Estonian acoustic models were trained on a 30 h set of broadcast news recordings [19]. The acoustic models were speaker-independent maximum likelihood-trained Hidden Markov models using Gaussian mixture models as emission probability distributions. The HMMs were state-tied triphone models with a gamma distribution for the state duration modelling.

The speech recognition experiments were performed in a broadcast news task for both Finnish and Estonian. For Finnish, the development set consisted of 5.38 h of audio with 35 439 word tokens and the evaluation set 5.58 h of audio with 37 169 word tokens. For Estonian, the development set consisted of 2.13 h of audio with 15 691 word tokens and the evaluation set 2.03 h of audio with 15 335 word tokens.

For earlier results on this speech recognition setup, see [17]. The Estonian results are comparable to the ones given in the subsection on decoding with subword units. For Finnish, one should note that we use the whole Kielipankki corpus for training, while in the larger setting of [17], the material was limited to one third of the corpus.

3.2 Language Models

All the evaluated language models were modified Kneser-Ney smoothed n-gram models [16] with three discounts per order [6]. The models were trained using the growing and pruning algorithm as implemented in the VariKN toolkit [27]. The training corpus was the same for all the models. A development set of 17000 sentences was used to optimize the discount parameters. The maximum n-gram order of the baseline word n-gram model was set to 3 because of the large vocabulary size. The growing parameter D was varied in the range 0.001–0.03 and the pruning parameter E set to twice the value of D to reach the model sizes in the Table 1.

For extracting the word classes for the class n-gram models, an optimized software with the word-class and class-word statistics [18] and multithreading [3] was used. The classes were initialized with a running modulo index over the word list sorted by decreasing frequency. This ensures that most common words are initially in different classes and speeds up the convergence. The algorithm was run for 24 h using 6 threads. The number of classes in the model was 1000 for both Finnish and Estonian. A 5-gram model was trained over the resulting class sequences.

The subword segmentation was trained by the method described in [36]. The method selects a subword vocabulary which codes the training corpus with high unigram likelihood. The subword n-gram models were trained over the resulting corpus segmentation. Special symbols were used for modelling the word boundaries. A 8-gram model was used for Finnish and a 6-gram model for Estonian. We trained two subword n-gram models, a smaller model to roughly match the number of parameters in the word n-gram and the second model to match the number of parameters in the interpolated word model.

3.3 Perplexities

We first report the perplexities of the language models. For Finnish, the evaluation was based on a held-out set of 347K sentences from the training corpus, which amounted to 4.1 million word tokens including sentence end symbols. For Estonian, the corresponding numbers are 200k sentences and 3.0M tokens. In the perplexity computations, the out-of-vocabulary rate for the word-based models was 2.3 % for Finnish and 0.9 % for Estonian. For subword models, the evaluation set was segmented with the corresponding subword n-gram model and the perplexity was normalized with the number of words. The interpolation weight in the linearly interpolated model was 0.6 for the word n-gram.

Table 1 shows the perplexity values. Given that the Finnish and Estonian languages are closely related, there is a large difference between the perplexity values. There are at least two possible reasons: First, the Finnish corpus included all types of newspaper articles and books, while the Estonian corpus was more focused on newswire style material. Second, the vocabulary size for Estonian is smaller and a slightly simpler preprocessing was applied. For both languages, the linear interpolation of the class n-gram model with the word n-gram model gave large improvements in perplexity. The relative decrease in perplexity was 21.3 % for Finnish and 12.6 % for Estonian.

Table 1. Perplexities for the language models. Both word and subword model perplexities are normalized per word, but they are not directly comparable due to different OOV rates.

Model	Finnish				Estonian			
	Vocabulary	OOV	Model size	PPL	Vocabulary	OOV	Model size	PPL
Subword n-gram	Unlimited	-	84.2M	2071	Unlimited	-	60.7M	627
Subword n-gram	Unlimited	-	110.0M	1912	Unlimited	-	97.6M	532
Class 5-gram	2.4M	2.3 %	32.3M	2194	1.6M	0.9 %	34.3M	883
Word 3-gram	2.4M	2.3 %	85.5M	1341	1.6M	0.9 %	59.3M	285
Word + class	2.4M	2.3 %	117.8M	1056	1.6M	0.9 %	93.6M	249

3.4   Speech Recognition Results

The language models were evaluated in the broadcast news task described in Sect. 3.1 for both Finnish and Estonian. The out-of-vocabulary rates for the word-based models in the evaluation set were 2.8 % for the Finnish task and 1.2 % for the Estonian task.

The results are in the Table 2. Note that the word error rates are high compared to what could be expected for example for English, as the words may be long and contain many derivational and inflectional suffixes. The baseline results with the unlimited vocabulary recognizers and the word n-gram are actually very good for this task [17]. The large vocabulary sizes are necessary to reach this recognition accuracy.

Table 2. Word error rates in a broadcast news task

Model	Finnish				Estonian			
	Vocabulary	OOV	Model size	WER	Vocabulary	OOV	Model size	WER
Subword n-gram	Unlimited	-	84.2M	30.0 %	Unlimited	-	60.7M	15.1 %
Subword n-gram	Unlimited	-	110.0M	29.8 %	Unlimited	-	97.6M	15.0 %
Class 5-gram	2.4M	2.8 %	32.3M	30.8 %	1.6M	1.2 %	34.3M	16.8 %
Word 3-gram	2.4M	2.8 %	85.5M	30.9 %	1.6M	1.2 %	59.3M	16.2 %
Word + class	2.4M	2.8 %	117.8M	29.2 %	1.6M	1.2 %	93.6M	15.0 %

The subword n-gram models provide good error rates, but the results improve very slowly by increasing the model sizes further. The word baseline results were 30.8 % WER for Finnish and 16.2 % WER for Estonian. The class n-gram model interpolation improved the WER by 5.5 % relative for Finnish and 7.4 % relative for Estonian. The results compared to the best subword recognizer are equal for Estonian and 2 % better for Finnish.

We note that the relative improvements in the perplexity and the word-error-rate differ between the languages. The evaluation data matched better to the acoustic and language model training data in Estonian than in Finnish. We tested also another setup for Finnish using data sets with a better match, and reached larger relative improvements in WER (around 10 %). However, the conclusions did not differ from the reported data sets.

The class n-gram models also performed surprisingly well as stand-alone models. For Finnish, the word error rate of the class n-gram was even slightly better than the error rate of the word n-gram although its perplexity value in Table 1 was considerably worse. By computing the perplexities on the broadcast news transcriptions, the perplexity difference was only 8.8 % relative in favor of the n-gram model. The absolute perplexity values were 2559 for the class n-gram model, 2333 for the n-gram model and 1709 for the interpolated model. This indicates that the class n-gram model was very robust with respect to slight mismatches in the data. Other reason may be related to sentence boundary modelling, as they have no effect for WER. It could also be that the class n-grams have good properties with respect to the speech recognition decision boundaries.

4   Discussion

In this work, class n-grams were studied for the very large vocabulary speech recognition of Finnish and Estonian. For morphologically rich languages like Finnish and Estonian, very high vocabulary sizes are required to achieve a reasonably low out-of-vocabulary rate in applications such as transcriptioning, dictation or broadcast news recognition. This will emphasize the data sparsity issues in training the language models. As the vocabulary size increases, it could be expected that the value of the different word clustering schemes and the language models trained over these classes will increase.

Training the classification with the exchange algorithm gave good results for both Finnish and Estonian. By interpolating the class n-gram model with the state-of-the-art modified Kneser-Ney smoothed trigram model, a 21.3 % relative reduction in the perplexity was reached for Finnish and a 12.6 % relative reduction in the perplexity for Estonian. Improvements in the perplexity are naturally dependent on the model sizes. For these experiments, fairly large baseline n-gram models were used. Different morphologically trained classes were also evaluated, but it was found out that the resulting models could be improved with the exchange algorithm.

Our current results with the class n-gram interpolation are equal for Estonian and 2 % better for Finnish when compared to an unlimited vocabulary subword recognizer. It is noteworthy that this happens solely by improving the language model estimates without attempting to model the OOV words. There may thus be possibilities for combining the modelling approaches, as they are at least to some extent complementary. It is unclear what language modelling methods would further improve the subword result. Class n-grams trained over subwords did not provide significant improvements at least with the word boundary modelling utilized in this work.

Different neural network language models have become popular in the recent years due to advances in the model training methods. Neural network language models for subword units have been evaluated for Finnish and Estonian LVCSR in [17]. However, it will be challenging to train the neural network language models over words for the vocabulary sizes studied in this work. Also, a good property of the class n-grams is that they may efficiently be applied in the first recognition pass without approximations. Thus, full advantage is taken from the improved modelling and also the lattice quality will be improved for possible further recognition passes.

The class n-gram approaches studied in this work should be applicable to other morphologically rich languages like Hungarian, Turkish, the Dravidian family of languages and others.

5   Conclusion

In this work, we evaluated class n-gram models for very large vocabulary speech recognition of Finnish and Estonian. Large improvements in perplexity were obtained by linearly interpolating with a standard word-based n-gram model. The models were compared to a state-of-the-art unlimited vocabulary speech recognizer using subword n-gram models. The results were comparable for Estonian and slightly better for Finnish. The stand-alone class n-gram model was very robust and performed for Finnish on par with the word-based n-gram model, with much reduced number of parameters.

Acknowledgments. This work was supported by the Academy of Finland with the grant 251170. Aalto Science-IT project provided computational resources for the work.

References

1. Aalto University: AaltoASR (2014). http://github.com/aalto-speech/AaltoASR/
2. Aubert, X.L.: An overview of decoding techniques for large vocabulary continuous speech recognition. Comput. Speech Lang. **16**(1), 89–114 (2002)
3. Botros, R., Irie, K., Sundermeyer, M., Ney, H.: On efficient training of word classes and their application to recurrent neural network language models. In: Proceedings of the INTERSPEECH, pp. 1443–1447, Dresden, Germany (2015)
4. Brown, P.F., deSouza, P.V., Mercer, R.L., Pietra, V.J.D., Lai, J.C.: Class-based n-gram models of natural language. Comput. Linguist. **18**(4), 467–470 (1992)

5. Brychcín, T., Konopik, M.: Morphological based language models for inflectional languages. In: The 6th IEEE International Conference on Intelligent Data Acquisition and Advanced Computing Systems: Technology and Applications, Prague, Czech Republic (2011)
6. Chen, S.F., Goodman, J.T.: An empirical study of smoothing techniques for language modeling. Technical report, TR-10-98. Computer Science Group, Harvard University (1998)
7. Creutz, M., Lagus, K.: Unsupervised discovery of morphemes. In: Proceedings of the ACL 2002 Workshop on Morphological and Phonological Learning. MPL 2002, vol. 6, pp. 21–30 (2002)
8. Creutz, M., Stolcke, A., Hirsimäki, T., Kurimo, M., Puurula, A., Pylkkönen, J., Siivola, V., Varjokallio, M., Arisoy, E., Saraçlar, M.: Morph-based speech recognition and modeling of out-of-vocabulary words across languages. ACM Trans. Speech Lang. Process. **5**(1), 1–29 (2007)
9. Deligne, S., Bimbot, F.: Inference of variable-length linguistic and acoustic units by multigrams. Speech Commun. **23**(3), 223–241 (1997)
10. Hirsimäki, T., Creutz, M., Siivola, V., Kurimo, M., Virpioja, S., Pylkkönen, J.: Unlimited vocabulary speech recognition with morph language models applied to Finnish. Comput. Speech Lang. **20**(4), 515–541 (2006)
11. Hirsimäki, T., Kurimo, M.: Decoder issues in unlimited Finnish speech recognition. In: Proceedings of the 6th Nordic Signal Processing Symposium (Norsig 2004), pp. 320–323, Espoo, Finland (2004)
12. Hirsimäki, T., Kurimo, M.: Analysing recognition errors in unlimited-vocabulary speech recognition. In: Proceedings of the HLT-NAACL, pp. 193–196 (2009)
13. Hirsimäki, T., Pylkkönen, J., Kurimo, M.: Importance of high-order n-gram models in morph-based speech recognition. IEEE Trans. Audio Speech Lang. Process. **17**(4), 724–732 (2009)
14. Iskra, D.J., Grosskopf, B., Marasek, K., van den Heuvel, H., Diehl, F., Kießling, A.: SPEECON - speech databases for consumer devices: database specification and validation. In: Proceedings of Third International Conference on Language Resources and Evaluation (LREC 2002), Canary Islands, Spain, May 2002
15. Kneser, R., Ney, H.: Forming word classes by statistical clustering for statistical language modelling. In: Proceedings of the First International Conference on Quantitative Linguistics (QUALICO), pp. 221–226, Trier, Germany (1991)
16. Kneser, R., Ney, H.: Improved backing-off for m-gram language modeling. In: Proceedings of the 1995 IEEE International Conference on Acoustics, Speech and Signal Processing (ICASSP), pp. 181–184 (1995)
17. Kurimo, M., Enarvi, S., Tilk, O., Varjokallio, M., Mansikkaniemi, A., Alumäe, T.: Modeling under-resourced languages for speech recognition. Lang. Res. Eval. 1–27 (2015)
18. Martin, S., Liermann, J., Ney, H.: Algorithms for bigram and trigram word clustering. Speech Commun. **24**, 19–37 (1998)
19. Meister, E., Meister, L., Metsvahi, R.: New speech corpora at IoC. In: XXVII Fonetiikan, 2012 – Phonetics Symposium 2012, pp. 30–33 (2012)
20. Mohri, M., Pereira, F.C.N., Riley, M.: Speech recognition with weighted finite state transducers. In: Benesty, J., Sondhi, M., Huang, Y. (eds.) Handbook on Speech Processing and Speech Communication, pp. 559–584. Springer, Heidelberg (2008)
21. Ney, H., Ortmanns, S.: Progress in dynamic programming search for LVCSR. Proc. IEEE **88**(8), 1224–1240 (2000)

22. Niesler, T., Whittaker, E., Woodland, P.: Comparison of part-of-speech and automatically derived category-based language models for speech recognition. In: Proceedings of the ICASSP, Seattle, USA (1998)

23. Niesler, T., Woodland, P.: Variable-length category n-gram language models. Comput. Speech Lang. **13**, 99–124 (1999)

24. Ortmanns, S., Ney, H.: Look-ahead techniques for fast beam search. Comput. Speech Lang. **14**(1), 15–32 (2000)

25. Pirinen, T.A.: Omorfi - free and open source morphological lexical database for Finnish. In: Proceedings of the 20th Nordic Conference of Computational Linguistics NODALIDA, Vilnius, Lithuania (2015)

26. Pylkkönen, J.: An efficient one-pass decoder for Finnish large vocabulary continuous speech recognition. In: Proceedings of the 2nd Baltic Confrence on Human Language Technologies (2005)

27. Siivola, V., Hirsimäki, T., Virpioja, S.: On growing and pruning Kneser-Ney smoothed n-gram models. IEEE Trans. Speech, Audio Lang. Process. **15**(5), 1617–1624 (2007)

28. Silfverberg, M., Ruokolainen, T., Lindén, K., Kurimo, M.: FinnPos: an open-source morphological tagging and lemmatization toolkit for Finnish. Lang. Resour. Eval. 1–16 (2015)

29. Sixtus, A., Ney, H.: From within-word model search to across-word model search in large vocabulary continuous speech recognition. Comput. Speech Lang. **16**(2), 245–271 (2002)

30. Soltau, H., Saon, G.: Dynamic network decoding revisited. In: IEEE Automatic Speech Recognition and Understanding Workshop, pp. 276–281 (2009)

31. Tarjan, B., Fegyó, T., Mihajlik, P.: A bilingual study on the prediction of morph-based improvement. In: Proceedings of the 4th International Workshop on Spoken Language Technologies for Under-resourced Languages SLTU, St. Petersburg, Russia (2014)

32. The Department of General Linguistics, University of Helsinki; The University of Eastern Finland; CSC - IT Center for Science Ltd

33. Vaičiūnas, A.: Statistical language models of Lithuanian and their application to very large vocabulary speech recognition. Summary of Doctoral dissertation. Vytautas Magnus University, Kaunas (2006)

34. Vaičiūnas, A., Kaminskas, V.: Statistical language models of Lithuanian based on word clustering and morphological decomposition. Inform. (Lith. Acad. Sci.) **15**, 565–580 (2004)

35. Varjokallio, M., Kurimo, M.: A word-level token-passing decoder for subword n-gram LVCSR. In: Proceedings of the IEEE Workshop on Spoken Language Technology, South Lake Tahoe, USA(2014)

36. Varjokallio, M., Kurimo, M., Virpioja, S.: Learning a subword vocabulary based on unigram likelihood. In: Proceedings of the IEEE Workshop on Automatic Speech Recognition and Understanding, Olomouc, Czech Republic (2013)

37. Whittaker, E., Woodland, P.: Efficient class-based language modelling for very large vocabularies. In: Proceedings of the IEEE International Conference on Acoustics, Speech and Signal Processing, Salt Lake City, USA (2001)

38. Whittaker, E., Woodland, P.: Language modelling for Russian and English using words and classes. Comput. Speech Lang. **17**, 87–104 (2003)

39. Young, S.J., Russell, N.H., Thornton, J.H.S.: Token passing: a simple conceptual model for connected speech recognition system. Technical report, Cambridge University Engineering Department (1989)

Author Index

Printed in the United States
By Bookmasters